FLIGHT TO MERCURY

Columbia University Press   New York

1977

BRUCE MURRAY and

ERIC BURGESS

Flight
to
Mercury

Bruce Murray is Director of the Jet Propulsion Laboratory of the California Institute of Technology.

Eric Burgess is a free-lance writer who has written extensively on the upper atmosphere, rockets, and space flight.

Library of Congress Cataloging in Publication Data

Murray, Bruce C.
  Flight to Mercury.
  Includes bibliographical references and index.
  1. Project Mariner. 2. Mercury probes. 3. Venus probes.
  I. Burgess, Eric, joint author. II. Title.
  TL789.8.U6M3496   629.43′54′1   76-25017   ISBN 0-231-03996-4

Columbia University Press, New York and Guildford, Surrey

To Christine, Stephen, Peter, and Allison
To Janis, Stephen, and Howard
And to all the other children of the Space Age

so that they may know a little more about their own extraordinary times

Preface

Spectacular exploits in space colored the 1960s and early 1970s and repeatedly captured America's attention. Indeed, few persons anywhere were indifferent to the televised views of astronauts—people like themselves—venturing onto the inhospitable surface of the Moon. And millions of people worldwide have been fascinated by close-up pictures of Earth's planetary neighbors, vistas that until our own times were beyond the reach of the most sophisticated science and eluded the most fantastic imagination. Today, space exploration has become an enduring part of the consciousness of Americans and of millions of other people throughout the world.

The many individuals involved in this historic endeavor are seen in much vaguer focus, however, than the unearthly landscapes being explored. The news process is necessarily a filter; its selectivity reinforces bland, antiseptic stereotypes in the public's mind. Even thoughtful, complex human beings such as astronauts can appear artificial and two-dimensional when viewed under the shadowless illumination usually provided for their exposure to the mass media. The technical engineers, scientists, and managers are even more indistinct. The challenging *human* endeavors of exploring beyond Earth are constructed from the successes and failures, passions and conflicts, of hundreds, sometimes thousands, of committed engineers, scientists, and managers. Their humanity is buried under hackneyed phrases such as: "Scientists at Caltech's Jet Propulsion Laboratory today. . . ." Their personal stories often exemplify many of the virtues that Americans value highly but somehow feel are disappearing from contemporary society. Yet these human stories are rarely told.[1]

---

[1] There are some conspicuous exceptions to this generalization, such as Henry Cooper's excellent *Thirteen—the Mission That Failed* (New York, Dial Press, 1973).

The professional jargon and specialized knowledge involved inhibit wide communication, even though the feelings, ambitions, and conflicts are the basic human attributes from which all drama is constructed. Moreover, often the participants are too busy, are not inclined toward introspection, or are inexperienced in popular writing. As a result, the panorama of individual dedication, imagination, and innovation in overcoming great obstacles for positive achievement—a vital aspect of current American life—goes largely unwritten.

Viewed from the future, this age will probably be remembered most for the emergence of mankind into space. Yet today this great uplifting human achievement is obscured beneath less broadly comprehensible science, technology, and economics.

We tell here the story of how and why mankind's first look at the innermost planet, Mercury, came about—and something of how it felt to be personally involved. Nevertheless, the portrayal is still muted and subdued compared to the sharp exposure to the original experience.

To communicate a little of what it is like to explore through robot eyes and detectors a distant and virtually unknown planet, the emphasis must be on a limited number of topics and events. Our story is therefore incomplete and subjective, and probably biased, but it is nevertheless an authentic personal view. For the most part, it records the efforts to obtain images of Mercury.[2] There were, of course, many other scientific objectives and experiments conducted at Venus and Mercury and during the voyage through interplanetary space. There were many other perspectives and feelings among the large number of scientists involved (their names and positions are given in Appendix A).

The Mariner 10 mission to Mercury took place during a particularly anguished period of American history. Unprecedented national and international events intruded upon the consciousness of participants as well as spectators of the Mariner 10 mission. Accordingly, the epoch-making headlines of 1973 and 1974 are used to punctuate the technical narrative of the flight itself as a contrapuntal reminder of the real world in which Americans struggled toward culturally valuable external achievements even as their national self-esteem was battered from within and without.

There were literally hundreds of people who made Mariner 10 happen. A few of them are leading characters in this story. More are listed in the appendices. The others are unfortunately too numerous to be cited here. But even the briefest of acknowledgments must emphasize that in

[2] The book reflects largely the personal views of Murray, an early proponent of the Mariner Venus/Mercury mission and eventual leader of the imaging team for it. Burgess has also compiled an official history for NASA, *The Voyage of Mariner 10* (in press), which synthesizes more impersonally a broader array of facts, opinions, scientific experiments, and experiences than does this book.

our democracy, dramatic, highly visible national enterprises like the Mariner 10 mission can only take place through genuine popular support. Mariner 10 and its accomplishment are lasting monuments to the optimism and imagination of the American people as they spearheaded mankind's emergence into the age of interplanetary exploration.

Acknowledgments

We greatly appreciate the efforts of James L. Anderson and Jurrie van der Woude of the California Institute of Technology in the design, layout, and preparation of the illustrations and captions for this book. Lorna Griffith has patiently prepared the many versions of the manuscript with her usual good humor and helpful comments.

Unless otherwise indicated in the captions, all figures and pictures are from the National Aeronautics and Space Administration. We have been particularly aided in obtaining appropriate versions of certain pictures by Audouin Dollfus (Fig. 1.1a); New Mexico State University Observatory (Figs. 1.1b, 4.1); S. Soter and J. Ulrichs from *Nature*, vol. 214, no. 5095, 1967 (Fig. 1.3); the Boeing Company (Fig. 2.7); Michael C. Malin (facing p. 33).

Contents

FLIGHT TO MERCURY

1

Prologue

50 Spruce Lane
Valley Stream N.Y.
November 18, 1974

Dear Mr. Murray

What have you found out about planet Mercury so far. How big is it? Do you think there has been any kind of life on Mercury? How far away is it from earth? How long have you been studying? Is it a hard job. How much do you know about it? Was that the only job you could find? I am 9 years old my birthday is October 14.

Sincerely
Tim Cavanaugh

P.S. Have a happy Thanksgiving!!!

Among the many revolutions during the last decade, one of the most important utilized space rockets instead of missile weapons, appealed to imagination and ingenuity rather than hatred and fear, and was characterized by discovery rather than destruction. It was accompanied by an explosion that shattered ideas rather than bodies.

During the last ten years, man—the first conscious creature in this solar system—reached out from his terrestrial birthplace to explore his planetary environment for the first time. A revolution in scientific ideas and popular perspectives about the neighboring planets, and Earth, has begun.

The extraordinary technological and social achievement of the Apollo project permitted astronauts to explore personally the lunar surface. Equally important to enlarging man's view of his planetary neighborhood have been the detailed pictures and scientific measurements returned to Earth from unmanned spacecraft sent close to Mars, Venus, and Mercury.

Since the advent of interplanetary spaceflight, the growth of information about the terrestrial planets, those rocky worlds that compose the inner solar system, has been stupendous. Suppose one tabulated all the useful Earth-based photographs of Mars laboriously acquired by astronomers over the half-century before 1965 (the year when Mariner 4 first signaled the opening of the space age for the Red Planet); they would barely equal the information contained in the 20 television frames sent back by radio from Mariner 4. In 1969, Mariners 6 and 7 flew by Mars and returned 100 times as much information as did Mariner 4. Mariner 9, which orbited Mars in 1971 and 1972, returned yet another 100 times as much information. The total photographic and other information about Mars thus increased 10,000-fold between 1965 and 1972! No wonder there has been a revolution in ideas about the nature and history of Mars during the last decade.

Knowledge of the Moon has increased even more explosively. Not only has there been an increase in photographic information about the Moon comparable to that about Mars, but also many sophisticated instruments such as seismometers have been placed on the lunar surface to operate there. Additionally, and most important, hundreds of pounds of carefully selected samples of lunar rocks and soil have been returned to Earth for study by the most sophisticated analytical techniques ever used.

Meanwhile, cloud-shrouded Venus was not neglected. In 1967, results from the American Mariner 5 flyby and the Soviet Venera 4 entry probe, combined with Earth-based radar observations, proved that Venus is shrouded by a very hot and dense atmosphere composed predominantly of carbon dioxide. Further Soviet flights have confirmed and elaborated on these results. Venus, Earth's twin in size and mass, proved to be most unearthly at its surface.

Of Earth's planetary relatives in the inner solar system, only Mercury— the innermost planet, which is not much larger than Earth's Moon— remained completely unexplored. Mariner 10, launched in November 1973, was intended to fill that gap. As a bonus, the trajectory of the spacecraft required a close pass of Venus on the way to Mercury, providing an opportunity for the first close-up photography of its mysterious cloudy atmosphere. Mariner 10 aimed to complete the initial exploration of the inner solar system, to finish the task of identifying common characteristics and heritage among Earth's planetary relatives. And despite many problems on Earth and in space, it succeeded beyond the hopes of its designers.

This is the story of that first flight to Mercury, of how it came to be and what it found. This story can be regarded as a case study from a revolutionary handbook—a cultural and intellectual revolution that further expanded human awareness beyond a global viewpoint.

Space exploration has significantly altered spatial and temporal perspectives of millions of people now living. The Earth-centered cosmology inherited from classical times was attacked by Copernicus over four centuries ago. Then Galileo provided experimental proofs of the Copernican alternative and also showed that the Moon had a mountainous surface. But it was not until American astronauts *walked* on that lunar surface, viewed through live TV by half a billion persons worldwide, that the Moon became a real, almost familiar place to a significant number of the world's populace. Similarly, pictures from unmanned Mariner spacecraft showed a strange and desolate but indisputably real Mars, thereby removing the Red Planet forever from the realm of speculative science fiction. These revolutionary changes in modern consciousness inevitably will propagate through succeeding generations to affect billions of people yet unborn. The Earth is not the center of the solar system and is no longer perceived as such by growing numbers of its inhabitants. Instead, it is seen as one of several rocky bodies, the inner planets, all of which are uninhabited and uninhabitable except the Earth.

There exists an apocryphal story that while that earlier revolutionary thinker, Nicolas Copernicus, lay on his deathbed in 1563, he bemoaned the fact that in his eventful life he had never seen the elusive innermost planet, Mercury. This is a doubtful anecdote since many northern Europeans reported sightings of the planet. But whether or not Copernicus ever did see Mercury, what was important for the study of Mercury and the other planets was the way his alternative theory of a Sun-centered solar system caused astronomers to observe the planets more accurately and to question "facts" that had been almost universally accepted. They focused attention on a place where Nature had important secrets to reveal as a result of careful observations—secrets that would have profound effects on the later course of the human species and would lead directly to the emergence of modern science and the technology of its application.

In 1973, five centuries after the birth of Copernicus, seven spacecraft were headed from Earth to explore the planets of the solar system. Mars—the next planet out from the Sun beyond Earth's orbit—was the target of four Soviet spacecraft designed to orbit the Red Planet and to

3

land capsules (unsuccessfully, as it turned out) on its windswept dusty surface. Beyond Mars the giant outer planets, Jupiter and Saturn, had two American spacecraft aimed toward them—Pioneer 10 and 11, each also destined to escape completely from the solar system and journey to the stars carrying a short message from mankind. The two innermost planets, Venus and Mercury, were being approached by Mariner 10, another American spacecraft.

Anyone could go out in the evening and see these planets in the darkening sky and marvel that in the five hundred years since Copernicus's birth mankind had moved from superstitious observers at a distance to curious close-up explorers. The planets were no longer bright points of light but places to be visited and explored. All the planets known to Copernicus were, indeed, now being visited by machines from Earth.

Today there is a great and expanding interest in the unexplained mysteries of the solar system, especially among the younger generation—that 50 percent of people on Earth today who were born during the space age. The exploration of the planets extends human senses over millions of miles and contributes to a new and deeper understanding of the Earth and its place in the universe, and to the role of mankind on the cosmic scale.

Although the Copernicus deathbed story is probably a fiction of some of his biographers, even the most experienced modern planetary astronomers have not viewed Mercury much better through a telescope than with the unaided eye. The small planet's great distance from Earth, its closeness to the Sun in the sky, and its narrowing crescent phase as it comes closer to Earth all make it a flickering enigma in even the best astronomical telescopes.

In 1800, in Lilienthal, Germany, Johann Schröter directed his 13-foot Schrader reflecting telescope at Mercury. He claimed he saw the southern horn of the crescent blunted. "It is a mountain eleven miles high," he announced. As he watched for several nights, he thought the appearance changed in a way that proved Mercury turned on its axis like the Earth—but in 24 hours and 4 minutes.

It was 80 years before another astronomer seriously looked at the innermost planet. In 1882 the director of the Brera Observatory in Milan, Giovanni Schiaparelli, peered at Mercury through the clearer skies of Italy during the daytime. After seven years of observations he announced, on December 8, 1889: "Mercury performs only one rotation during its revolution round the Sun." Theoretical astronomers found this observation plausible. George Darwin's tidal theory was at that time one of the most striking theoretical developments in solar system astronomy, explaining why the Moon always turned one hemisphere toward Earth. Mercury's proximity to the Sun could cause synchronous rotation also.

Astronomers accepted this "fact" for many years; a myth of Mercury gained credence. Nearly everyone agreed that the planet must rotate on its axis synchronously with its revolution around the Sun, and thus must always turn one face toward the Sun, as the Moon turns one face toward the Earth. As a consequence, one side of Mercury would have everlasting day and the other, everlasting night.

The myths about Mercury dominated popular and scientific viewpoints for almost a century. People talked about atmosphere frozen solid on the dark side of the planet, "canals" like those on Mars, mountains, sheets of water, volcanoes in action, and an atmosphere; but a continued argument raged over the period of rotation. One observer claimed a short rotation determined to within 0.8 seconds; others were adamant about the rotation being synchronous with revolution. Much of this mass of "fact" was merely speculation that bore little resemblance to what later proved to be the truth.

As recently as 1962 a leading expert in visual and photographic planetary observations wrote, in the most authoritative book on the planets at that time, that Mercury's rotation was synchronous with the Sun to one part in ten thousand. Also, a thin atmosphere on Mercury was inferred by some spectroscopists as well as by another astronomer observing the variation of polarization of light reflected across the disc of the planet. Only heavy gases could be retained by Mercury to form any tangible atmosphere because of its low gravity and high temperature.

Some astronomers considered that Mercury's surface would be Moon-like, and science fiction artists painted the surface of Mercury as very similar to that of the atmosphereless Moon. This concept was suggested by indirect astronomical observations—the way the planet's surface affects radio waves and emits infrared radiation—though it was impossible to see other than faint markings on Mercury from Earth (Fig. 1.1).

The highly inclined orbit of Mercury made it necessary for the ancients to extend the width of the Zodiac (the band of the celestial sphere along which the planets are observed) to 16 degrees. Changes in the movement of Mercury's perihelion (the point on its very eccentric orbit where the planet approaches closest to the Sun) of about 40 seconds of arc per century caused some early disbelief in Isaac Newton's inverse square law for gravity. Simon Newcomb suggested that on the basis of Mercury's orbital anomalies, the exponent of the distance in the law of gravity should be 2.0000001574 instead of 2 precisely. Others, such as Urbain Leverrier, speculated that there must be a perturbing planet closer to the Sun. A number of searches were made, all unsuccessfully, for this hypothetical planet, Vulcan. It was only in 1915 that the anomalies of Mercury's orbit were explained by Albert Einstein as a consequence of general relativity.

5

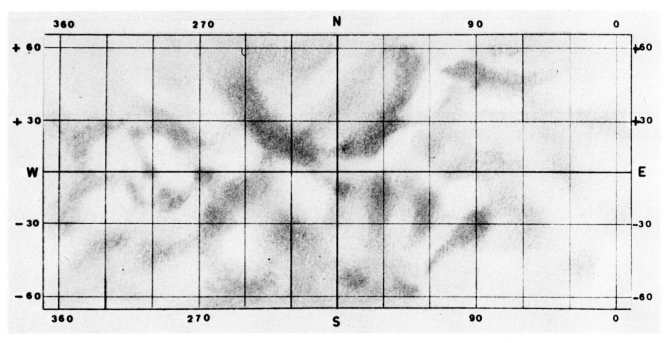

Figure 1.1a   Maps of Mercury prior to Mariner 10, such as this one compiled by A. Dollfus from 130 selected drawings of the planet, gave few clues to the true nature of the surface.

The reflection of sunlight and radar waves and the emission of thermally generated infrared and radio waves indicated that the surface of Mercury closely resembled what the Moon would look like if it were located at the same distance from the Sun as Mercury. This similarity to the Moon in surface materials presented problems for interpretation of the mean densities of the terrestrial planets on a common basis. Astronomers had known for many years that Mercury is a dense planet, much denser than the Moon and Mars and rivaling the Earth. The bulk or average density of the Earth as a whole is greater than the average density of material near its surface because gravity compresses the materials near its center. A similar bulk density for the smaller and less massive Mercury means that there must be a greater concentration of heavier elements (especially iron) in Mercury than in Earth, quite unlike the Moon.

The major question was whether Mercury is a homogeneous mix of iron and silicates (like some of the meteorites) or has a large iron-rich core surrounded by a mantle of lighter materials (like the Earth). If the latter were true, then the core would be three-fourths the diameter of the planet, about the size of the whole Moon! By comparison, Earth's core is just over half its diameter (54 percent).

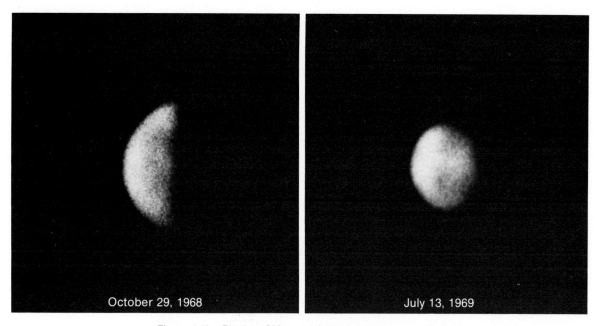

October 29, 1968   July 13, 1969

Figure 1.1b   Photos of Mercury through telescopes on Earth show only faint dusky markings. (Photo: New Mexico State University Observatory)

Mercury was first explored by radio telescope in 1962 by a group from the University of Michigan, observing the planet when it appeared in the telescope with "half-moon" illumination. They sought to detect radio waves emitted by the hot surface of Mercury's darkened hemisphere. It was expected that, if Mercury did not rotate rapidly but continually faced the same hemisphere toward the Sun, the dark, shadowed side would be eternally shaded and would therefore be extremely cold. Such a cold surface would not emit radio signals detectable with the instruments used. The University of Michigan researchers were surprised to find "excess" thermal radio waves coming from the dark parts of the disc of Mercury. Was there actually warm soil radiating radio waves from the night hemisphere of Mercury? From the radio standpoint, Mercury was behaving as the Moon would at the distance of Mercury—but rotating slowly on its axis and not in synchronism with its revolution around the Sun. However, so deeply ingrained was the idea of synchronism of rotation with orbital revolution that *no one* considered that other possibility—fractionally synchronous rotation. Instead, astronomers searched for evidence of an atmosphere necessary to carry heat from the day to nighttime surfaces and thereby explain the radio observations.

7

Finally, in 1965, Rolf Dyce and Gordon Pettingill looked at Mercury by radar from the giant antenna at Arecibo, Puerto Rico. They bounced radio waves off Mercury, and from the returned echoes they concluded that Mercury does not rotate synchronously about the Sun (i.e., 88 days) but in a period of about 59 days.

This unexpected result led an Italian astronomer, Guiseppe Colombo, of the University of Padua, to recognize that the observed radar rotation period was close to being in a 3:2 ratio with the 88-day orbital period. Mercury, he conjectured, was rotating exactly three times for every two revolutions around the Sun (Fig. 1.2). More precise radar observations later proved this speculation to be correct. A theory has now been developed to show how Mercury could have been slowed down to this 3/2 spin-orbit coupling as a consequence of great solar tidal effects. It was, indeed, this unusual coupling between the spin and the orbital revolutions that accounted for mistaken reports about Mercury, suggesting a synchronism of rotation and revolution. Mercury can never be observed effectively from Earth around a complete orbit. Because every two revolutions the planet does present the same face to the Sun, and this coincides with opportune times to observe Mercury from Earth, it is easy to see how earlier optical observers of this difficult planet could fall into a trap.

Mercury has the harshest surface environment in the solar system. At perihelion the solar radiation falling on Mercury is 10 times that on the Moon (Fig. 1.3). At the equator of Mercury at noon the surface reaches 800°F (427°C), while the nighttime-hemisphere temperature plunges to below −280°F (−173°C). The perihelion noon lasts several weeks because the speed of Mercury along its orbit is fastest at perihelion while the spin angular velocity does not change. An observer on Mercury would see the Sun stop its movement across the sky and go backward for a short while. Or if located at the terminator, the boundary between night and day, the observer would witness a double sunrise or a double sunset. Exciting prospects for future astronauts!

The peculiar spin-orbit coupling also brings first one area and then another diametrically opposite or antipodal area directly under the Sun at perihelion. These are veritable hot spots on the planet where the greatest temperatures are experienced, receiving 2½ times as much solar radiation as similar areas that are 90° in longitude away.

By 1973, little was really known about the innermost planet except some basics—its size, mass, density, orbit, rotation period, and the fact that its surface was covered with powdered silicate material similar to that found on the surface of the Moon. But the surface landforms and internal constitution of Mercury were a mystery. It was not even known whether Mercury possessed a thin atmosphere. In many ways, we knew

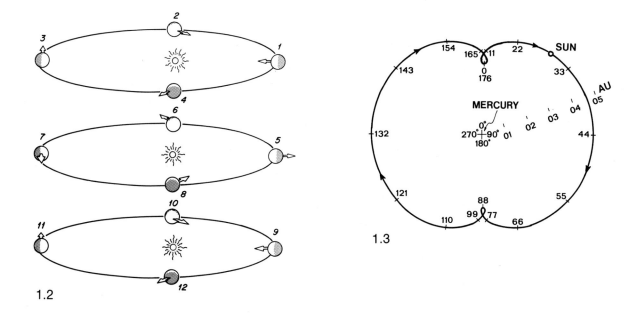

1.2

1.3

Figure 1.2 Mercury orbits the Sun in 88 days and rotates on its axis in 58.65 Earth days. This 3/2 spin-orbit coupling results in Mercury's rotating precisely three times on its axis for every two revolutions around the Sun. A solar day on Mercury is thus 176 Earth days.

Figure 1.3 As seen from Mercury, the apparent path of the Sun produces two loops at perihelion, when for a short period the Sun halts and moves backward in the sky. This retrograde motion causes the Sun to rise and set twice each day on some parts of Mercury. The points of the surface facing the Sun at perihelion therefore receive more solar radiation and reach higher noontime temperatures than elsewhere on the planet. (After S. Soter and J. Ulrichs, *Nature*, 1967)

little more about the moth-like companion of the Sun than did Copernicus. Mercury had merely changed from a point of light tracing out a complex path near the Sun to a 3,050-mile (4,860 km) diameter, indistinct, heavy ball, spinning on its axis three times for every two revolutions around the Sun in an orbit that wobbles like a hula hoop at an average distance of 36,000,000 miles (58,000,000 km).

However, there arose a unique opportunity to explore Mercury. The United States had developed some unprecedented technological skills, notably in the art and science of commanding and controlling space vehicles with great precision over long interplanetary distances, and even more important, of being able to send information back from such spacecraft at very high rates and with great reliability.

At the same time the planets themselves cooperated by presenting a special configuration that would enable the spacecraft mission planners to design a mission to Mercury using the gravity and orbital motion of Venus to assist the spacecraft and thereby allow a Mariner type of craft to be sent all the way to Mercury with only the regular launch vehicle needed for a Mars or Venus mission.

Thus, 1973 emerged as an opportune time for an attempt to address basic questions about the innermost world. What kind of planet is Mercury? What is its surface like? Is its surface cratered like the Moon or crumpled like Earth? Does it have a large, iron-rich core enclosed in a silicate mantle and crust (like a miniature Earth)? What has been the planet's history, and what does that tell us about the history of Earth?

The time had come for another revolutionary step: Man's first close-up look at planet Mercury.

9

2

Two Stones with One Bird: Birth Pangs

"Providing the time of departure is carefully arranged, Mars and Mercury will be seen at close range, while a near approach to Venus will be made at least once, perhaps twice."
—Phillip E. Cleator, in *Rockets Through Space*, 1936

The exploration of Mercury was made possible by an exciting new idea in astronautics: the use of the gravitational attraction and orbital motion of Venus as a celestial slingshot to hurl a spacecraft sunward to intersect the orbit of Mercury. Without this technique a flight to Mercury could not be made with a conventional Mariner-class launch vehicle. Much larger systems had been developed to send astronauts into space and ultimately to the Moon in project Apollo, but they were considered far too costly for unmanned missions to the planets.

A dual-planet mission, actually a multi-planet mission, had been discussed as long ago as 1936 by Phillip Cleator. But almost 20 years passed before the real power of the multi-planet mission, the gravity-assist slingshot to conserve propellant mass, was mentioned again. In 1954, Derek F. Lawden said, in the *Journal of the British Interplanetary Society:*

> [A] velocity increment [could be] induced in a space ship due to its attraction by a large moving body and without expenditure of fuel . . . a perturbation maneuver is seen as a means of economizing in the fuel requirements of an interplanetary journey.

But the technique remained virtually forgotten for another decade until the rocket rivalry of the United States and the Soviet Union began spreading beyond Earth to other worlds.

11

1961	Days of excitement and activity in the space race. Students, encouraged and financed through summer programs, work at aerospace companies and National Aeronautics and Space Administration (NASA) centers, excited by the challenge of this new frontier. Michael A. Minovitch, a graduate student at the University of California at Los Angeles, spends several summers at the Jet Propulsion Laboratory, Pasadena, California. His task: to calculate a series of Earth–Venus–Earth and other possible round trip missions.
October, 1962	Minovitch's work at the Laboratory culminates in the discovery that in 1970 and 1973 a spacecraft launched to Venus could be perturbed by a close passage into a new trajectory that would encounter Mercury.

By early 1967, the basic feasibility of the mission to Mercury via Venus has been established and its scientific and exploratory potential recognized. Furthermore, it is apparent that after the planetary configurations of 1970 and 1973, there will not be another practical opportunity for an economical voyage to Mercury until the middle 1980s. But the deadline to start work on any 1970 effort passes almost unnoticed, except by a few disappointed scientists.

June, 1968	Good news. The Space Science Board of the National Academy of Sciences (NAS), after a study on planetary exploration, endorses a mission to Mercury at the 1973 opportunity:

> Our emphasis on obtaining broad knowledge of the solar system leads us to give the next [i.e., after a Mars orbiter] priority to a Mariner-class Venus/Mercury flyby in 1973 or 1975.

Shortly after the NAS recommendation, the objectives of such a mission are defined by NASA, and serious engineering studies are begun. In addition to providing the first look at Mercury and new observations of Venus, the spacecraft would explore the interplanetary medium within the orbit of Venus for the first time. Moreover, the Venus/Mercury flight would provide experience in the gravity-assist technique that will also be needed for later exploration of the outer giants of the solar system—Jupiter, Saturn, Uranus, and Neptune.

September, 1969	NASA selects a representative group of scientists to participate in planning the Venus/Mercury mission—a group that becomes the Science Steering Group (SSG). Such early involvement of scientists is an innovative approach by NASA for Mariner spacecraft. In the past, these spacecraft had been specified and designed before scientific experiments were solicited, a practice that in some cases had placed severe constraints on the experiments.

Meanwhile, support for the mission to Mercury grows steadily. The Lunar and Planetary Missions Board, consisting of senior scientists from outside of NASA, strongly recommends the mission:

After reviewing the planetary program for the next five years, the Board feels strongly that the Mercury/Venus flyby in 1973 represents an important part of the whole program. Mercury, because of its anomalously high density, nearness to the Sun, peculiar revolution and rotation characteristics, and unknown atmosphere, is of importance in studies of the origin of the planets and the solar system. Our only practical opportunity, for many years to come, to deliver a spacecraft to the vicinity of Mercury is to use a swingby past Venus in 1973. This will advance the technology for which there will be great use in the further investigation of far distant planets.

Fall, 1969   NASA headquarters reviews the proposed mission and finds the prospects encouraging. The mission is submitted to Congress by NASA's Office of Space Science and Applications as a new start for fiscal year 1970, with a proposed funding level of $3 million.

A major setback immediately follows.

Although the Venus/Mercury project is never challenged as a worthwhile exploration of the planets, the House Committee on Space Science and Applications recommends that the project be deferred so that more money can be allocated for Earth-orbital applications instead of planetary exploration. "Defer" is, for this mission, "cancel," because the planetary configuration needed for the mission does not repeat until the 1980s. Mariner Venus/Mercury suddenly is caught between NASA and the Congress in a tug-of-war over priorities in space. The great opportunity for United States science and technology to carry out a historic accomplishment seems destined to founder on the shoals of political confrontation.

The Senate supports the project, however, and includes in its version of the authorization bill the $3 million required for the first year's work.

November 18, 1969   The House–Senate Authorization Conference Committee approves the mission for fiscal year 1970 along with an additional $10 million for Earth-orbital applications. Confrontation yields to compromise.

The Jet Propulsion Laboratory (JPL) of the California Institute of Technology has been highly successful with Mariner spacecraft in exploring both Mars and Venus. For the Venus/Mercury mission the Laboratory proposes a Mariner-type spacecraft to be launched by an Atlas/Centaur rocket. It plans to draw heavily on subsystems left over from the earlier Mars missions.

A new challenge then appears. NASA insists that this Venus/Mercury mission must be the first of a new breed of low-cost missions. It must establish a new standard for economy and cost control in planetary missions, yet still penetrate and operate in a more hostile physical environment than any previous spacecraft.

13

December 15, 1969   Dr. William Pickering, Director of JPL, takes the initiative by writing a letter to NASA, committing the Laboratory to conduct the Mariner Venus/Mercury 1973 program within the NASA guidelines for a fixed cost of $98 million, regardless of environmental and technical uncertainties. With looming economic problems, an increasing rate of inflation, and an as-yet-untried gravity slingshot technique to reach and explore an almost completely unknown planet, this letter from Dr. Pickering shows great confidence in the ability of Laboratory personnel to overcome unforeseen difficulties without requesting extra money from NASA. This is the first planetary mission to be carried out at a fixed price. Space exploration has clearly matured from early days when Ranger spacecraft failed in the first attempts to reach the Moon. Now, a decade later, the space arena has been confidently extended from the thousands of miles of cislunar space (between Earth and the Moon) to the millions of miles of interplanetary space.

NASA is quick to accept the JPL proposal. In January 1970 a Mariner Venus/Mercury project office is established at JPL. Although man's first look at the innermost planet has finally been conceived, a difficult period of 45 months gestation lies ahead before the spacecraft can be "born" and carried into space.

Meanwhile, the Science Steering Group of NASA has developed scientific objectives for the mission. For the Mercury encounter the SSG recommends that images be obtained to identify the general physical nature of the planet and to provide details of surface features and how they reflect the sunlight illuminating them. The mission should also refine Earth-based information about the shape and size of Mercury, its period of rotation, and its axial tilt. Other experiments should measure the surface temperature and try to find out if Mercury possesses an atmosphere and, if so, to determine its constituents. The SSG also recommends experiments to find out how Mercury interacts with the solar wind—an invisible blizzard of electrons, protons, and other particles emanating from the Sun. Such investigations can indirectly reveal the presence of a planetary magnetic field, although Mercury is not expected to exhibit a significant one because it rotates so slowly on its axis. The only other planets known to have significant magnetic fields (Earth and Jupiter) both rotate rapidly.

At the Venus flyby, the imaging system should investigate whether there is any fine structural detail visible at the top of the Venus clouds and should record at high resolution the faint ultraviolet markings photographed from Earth and how these change over several days. No previous spacecraft has photographed Venus.

Since the spacecraft necessarily will pass behind Venus as seen from Earth and thus be occulted, or hidden from view, radio signals can be used (as in earlier Mariner spacecraft) to probe the atmosphere of Venus. The changes in the path of the spacecraft caused by the gravitational field of Venus and measured very precisely by observations of the radio signals from the spacecraft will be used to improve existing information about the planet's mass, shape, and gravity.

The SSG also has recommended a trajectory for the flyby of Mercury—a crucial issue. The spacecraft is to make a darkside, near-equatorial flyby during which both the Sun and the Earth would be occulted by the planet as seen from Mariner (Fig. 2.1). Solar occultation will take the spacecraft through the wake of Mercury to look for an expected cavity in the solar wind like that behind the Moon. Earth occultation will allow

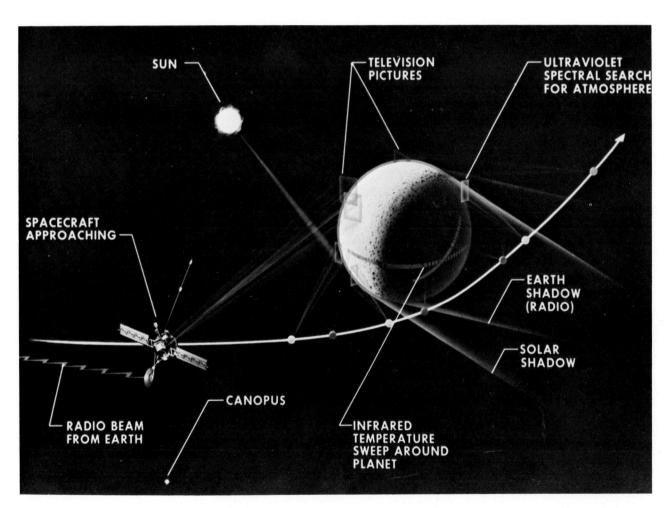

Figure 2.1 The encounter geometry chosen for Mercury was clearly unfavorable for photographic imaging, passing by the planet on its dark side. This choice was mandated, however, by the necessity of passing both through the Sun's shadow and behind Mercury as seen from Earth to obtain vital measurements of the small planet's interaction with the solar wind and of any tenuous atmosphere it might have.

15

radio waves from the spacecraft to graze the limb (edge of the planet's disc) of Mercury to detect any atmosphere or ionosphere and refine our measurements of the physical dimensions of the planet.

To perform all these science experiments it is planned that Mariner Venus/Mercury will carry a battery of scientific instruments (see Fig. 2.4): the magnetometer measures magnetic fields; the plasma analyzer measures the ions and electrons flowing through space from the Sun (the solar wind); cosmic ray telescopes record the solar and galactic cosmic rays. These experiments are all directed toward ascertaining how the planet interacts with the interplanetary environment of magnetic fields and energetic particles.

An infrared radiometer (heat-measuring device) will measure the temperature of the clouds of Venus and the surface of Mercury. Two independent ultraviolet instruments, which measure light beyond the visible region of the spectrum, are intended to analyze the composition of the planetary atmospheres. One of these ultraviolet instruments is fixed to the body of the spacecraft and will be used at Mercury to search for traces of a rarefied atmosphere along the edge of the planet's disc. The other is mounted on the movable platform that holds the TV cameras and can be pointed on command. This second instrument will be used to scan the entire discs of both planets, searching for traces of hydrogen, helium, argon, neon, oxygen, and carbon evaporating into space. Close to Earth this same instrument will measure the hydrogen corona of the Earth and the reflective properties of the lunar surface for comparison with Venus and Mercury respectively.

Television cameras attached to telescopes will be used to send back pictures of the planets, and the spacecraft's radio signals will be used to determine how the gravitational fields of the planets affect the path of the spacecraft. This provides information about the mass and shape of the planets. Additionally, during occultations, the interruption of the radio signals provides further information about atmospheres and planetary dimensions.

Although the nightside pass provides the best opportunity to investigate the interaction of Mercury with the solar wind, to study the heat radiated from the planet's surface, and to search for a very tenuous atmosphere, it is the worst trajectory for photographing the planet's illuminated surface: when the spacecraft is closest to the surface, it will be over the night hemisphere. As a compromise, the imaging team is permitted to design and develop a new optical system of threefold higher resolution than previous cameras used on Mariner missions to Mars. The new camera system will provide high-resolution pictures from the much greater distances required by the darkside passage.

At a conference at the California Institute of Technology, a large group of scientists are asked to consider the implications of the SSG recommendations. Much interest is created. The scientific exploration of Mercury is beginning to sound plausible. It was at this conference that a novel idea for extending the mission was broached.

"Dr. Murray, Dr. Murray. . . . Before I return to Italy, there is something I must ask you."

"What is it, Dr. Colombo?"

"Dr. Murray. . . . What will be the period of the spacecraft about the Sun after the Mercury encounter? Can the spacecraft be made to come back?"

"Come back?"

"Yes, the spacecraft could return to Mercury."

"Are you sure?"

"Why don't you check?"

Dr. Giuseppe Colombo, the Italian astronomer who recognized the 3/2 spin-orbit resonance of Mercury, explains his idea. After the spacecraft flies by Mercury, it will continue to orbit the Sun with a period very close to twice that of Mercury. Thus the spacecraft will travel once around the Sun while Mercury makes two revolutions on its orbit. Six months after the first encounter with Mercury, the spacecraft will approach the planet again for a second encounter, and so on. Murray is intrigued at the thought of the extended mission:

"Well, I'll ask the navigation engineers at JPL to run out the trajectory on the computer to see what happens after Mercury encounter."

The extended mission was, indeed, feasible. The spacecraft would return to Mercury. But the big question was whether or not the spacecraft could survive the additional flight time required for these subsequent encounters. Would there be sufficient maneuvering gas and propellant for the necessary trajectory corrections to ensure the second and third encounters?

Because Mercury will rotate on its axis three times in the six months between encounters, the spacecraft will view the same hemisphere of the planet illuminated at each encounter. Whereas Mercury itself has a 3/2 spin-orbit coupling, the spacecraft and Mercury will have a 1/2 resonance in their orbits around the Sun.

The navigation and trajectory concepts are now clear. Only the United States appears to have the capability for such a precision mission, one in which any uncorrected error in aiming at Venus will be translated into an error 1,000 times greater at Mercury. Unless the flyby of Venus can be kept to very close tolerances of distance and time, the spacecraft will pass Mercury at too great a distance to take useful pictures or to make other scientific measurements.

Additionally, because of the rapid passage of the spacecraft past Mercury, sophisticated telecommunications will be needed to obtain television picture coverage of a large area of the planet at high resolution. The resolution of a picture is a measure of the fineness of the details recorded. The American technology to do this is far in advance of any other nation's and therefore the scientists and engineers are confident that hard work, experience, and ingenuity will suffice. Yet, as the program advances toward the launch date, it becomes clear that even these unique technical capabilities and personal optimism are by no means sufficient to get the required imaging data from the distance of Mercury in the short time of the flyby over the dark hemisphere if Mariner Venus/Mercury has to use the same equipment and techniques as those used for earlier missions to Mars.

This problem of getting sufficient information back from the spacecraft gradually develops into a full-fledged "battle of the bits." Information gathered at a distant planet by a spacecraft, including photo-images of the planet, is returned to Earth over radio communication links in a form roughly analogous to the dots and dashes of the Morse code. The picture is broken down into strips and each strip is, in turn, broken down into individual elements, or pixels. Thus the picture is changed into a long sequence of signals which at the ground station is converted back into picture strips and then into the complete picture. The higher the rate at which the "bits" of information making up the individual pixels of the picture can be sent to Earth, the more detailed the information that can be transmitted. Television imaging of the surface of a planet requires the highest data rate of all the scientific experiments on a planetary mission.

The number of individual pictures that can be taken at a certain level of detail and sent to Earth during the time of the proposed Mercury flyby is limited by the communications rate. To meet even the basic photographic objectives, improved communications capability is essential for this mission to Mercury.

Early 1970   Direct application of previous systems and procedures used on missions to Mars will limit the communications rate to 12,000 bits per second from Mercury—referred to by the engineers as 12 kilobits of data. This is only one-tenth the rate necessary for a first-class imaging experiment (Fig. 2.2). The data-rate drama begins. Can a *tenfold* increase in communications capability be created, within the unprecedented fixed-price guidelines?

Meanwhile, the spacecraft itself must be designed, constructed, tested. NASA decides that Mariner Venus/Mercury will be designed, built, and

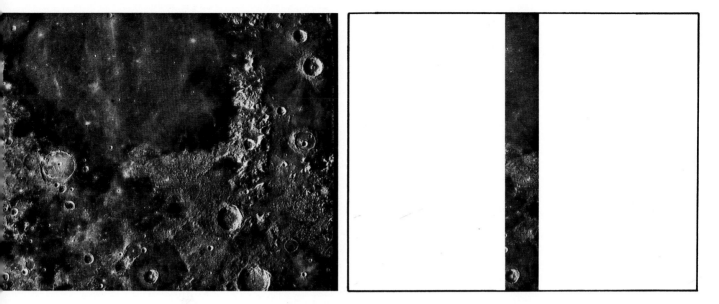

Figure 2.2 The desired photo-coverage at Mercury is illustrated on the left, using a picture of the Moon for an example. The communications capability initially considered in early 1970 for the Mariner 10 mission was equivalent to the right-hand side, about one-tenth of that needed. Note that the preferred coverage not only shows different terrains but also adds to the interpretability of the center by revealing its setting.

tested through outside contracting, with JPL providing project management. The decision to work with a system contractor rather than produce the Mariner Venus/Mercury spacecraft within the Laboratory as with earlier Mariners causes much soul-searching by JPL management. The Laboratory's experience with working through a contractor for a spacecraft on a major program is limited to the Surveyor project to softland a spacecraft on the Moon in the early 1960s. That episode was characterized by contracting difficulties, technical problems, delays, cost overruns, and sharp Congressional criticism.

Fortunately, the Surveyor project manager who helped to retrieve that program from serious problems is available again. Often referred to as an incurable optimist with an uncanny ability to get the job done, Walker E. (Gene) Giberson is appointed Project Manager for Mariner Venus/Mercury in January 1970. His first activity is to select a small team who have demonstrated project ability as well as capability of working within tight budgets: John Casani for the spacecraft, Victor Clarke for mission analysis and engineering, and Dr. James Dunne as project scientist. Almost immediately Vic Clarke becomes the champion of increased capability and performance of the still hypothetical spacecraft—and Casani inevitably is cast in the role of the tough-minded and

19

practical manager who has to protect the project from the once-inevitable cost increases resulting from changes and "improvements." For openers, Vic Clarke proposes that a higher communications rate be obtained by using a 20-watt X-band transmitter instead of the 20-watt S-band system of the previous Mars missions. Sections of the radio frequency spectrum are referred to as bands. The international allocation is S-band, from 2290 to 2300 megahertz (MHz), and X-band, from 8400 to 8500 MHz. Each space mission uses only small channels within these bands. The higher bands permit more information to be transmitted per second. All other things being equal, use of the higher frequency X-band would permit a 13-fold increase in communication rate. Project management doubts whether an X-band system can be made reliable without excessive cost, so this proposal is rejected.

Fall, 1970   Another scene in the data-rate drama: pressure from the imaging team skillfully modulated by Vic Clarke leads to reconsideration of using the higher radio frequency (X-band) to send back the TV pictures. At the December design review, a diagram of an X-band system is shown that would provide the needed communication rate—117,000 bits per second (117 kilobits) at Mercury. Vic Clarke announces dramatically that he has sought and found the necessary equipment. But X-band has not been proven in previous interplanetary work, and highest possible reliability is mandatory. The X-band is again rejected.

February 1, 1971   Four contractors submit proposals to JPL for the spacecraft, based on specifications calling for only limited communications. Although full communication capacity of 117 kilobits is planned for Venus, using a 48-inch diameter antenna to return the pictures, only 22 kilobits is planned from Mercury. But a backup capability of 44 kilobits from Mercury is held as a possibility. It would improve the photo coverage (Fig. 2.3).

June 17, 1971
Seattle, Washington   The Boeing Aerospace Company, successful bidder on the spacecraft, starts work on design and development. Only one flight spacecraft is planned.

And in testimony before the House Space Committee hearings:

"You may continue your testimony, Dr. Murray."

"Thank you, Mr. Chairman. It should not be overlooked that Mariner Venus/Mercury, potentially our most cost-effective and competitive planetary mission, is based on a single launch with no backup plan whatsoever. If you want to use terms that a businessman might use concerning capital investment, the risk of an Atlas/Centaur launch, based on several years experience, is about one failure in four. So we're investing $98 million in a spacecraft on a three-out-of-four probability that the spacecraft will even get out of Earth orbit."

The Mariner Venus/Mercury is to be the fifth of the series that began with Mariner Mars '64, followed by Mariner Venus '67, Mariner Mars

20

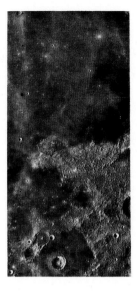

Figure 2.3 Fraction of lunar frame simulating reception of 44 kilobit data rate from Mercury, included as a possible rate if conditions were favorable.

'69, and Mariner Mars '71. In common with these earlier spacecraft, the new Mariner is designed around an octagonal main structure with eight equipment bays, a monopropellant propulsion system for trajectory corrections, solar cells and a battery for electrical power, a stabilization system of gyros and nitrogen gas thrusters for orienting the spacecraft, star and sun sensors to determine its orientation, and a movable platform to point the telescopes of the imaging system (Fig. 2.4).

Changes are required, however, because no Mariner spacecraft yet designed can penetrate into the inner solar system without modification. It will be entering an environment where the intensity of solar radiation reaches almost five times that at Earth's orbit. The spacecraft itself needs added protection. Also, the solar cells used to convert solar radiation into electrical power must be prevented from overheating.

To keep the solar cells at the right temperature, the panels on which they are mounted have to be tilted away from the Sun. Two panels are planned for Mariner, extending on either side of the spacecraft like sails. The designers have to decide how to tilt them away from the Sun: by raising them both upward to form a V shape, or by rotating each one along its long axis. The latter way would allow attitude-control thrusters to be mounted at the end of the solar cell panels in positions that would conserve their use of nitrogen gas. Also, the V configuration has undesirable thermal coupling between the body of the spacecraft and the solar panels which aggravates the problem of keeping the spacecraft cool. The rotating panel configuration is chosen.

21

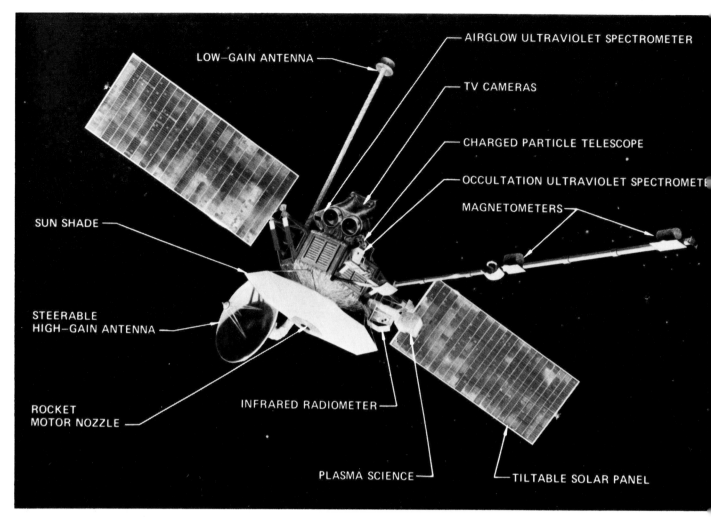

LOW–GAIN ANTENNA

AIRGLOW ULTRAVIOLET SPECTROMETER

TV CAMERAS

CHARGED PARTICLE TELESCOPE

OCCULTATION ULTRAVIOLET SPECTROMETE

MAGNETOMETERS

SUN SHADE

STEERABLE
HIGH–GAIN ANTENNA

ROCKET
MOTOR NOZZLE

INFRARED RADIOMETER

PLASMA SCIENCE

TILTABLE SOLAR PANEL

Figure 2.4 The Mariner Venus/Mercury spacecraft carried a battery of scientific instruments to explore the inner planets.

Although not anticipated at the time it was made, this decision proves invaluable later in the mission when it becomes necessary to use the rotatable panels as solar sails to control the orientation of the spacecraft when the maneuvering gas supply becomes low.

The path chosen for the spacecraft to fly by Mercury over the night hemisphere causes problems in imaging which, in turn, require modifications to the imaging and data systems of the spacecraft. Because a nightside pass has been chosen, photographs cannot be taken at closest approach to Mercury but only from greater distances as the spacecraft travels toward and away from the planet. As mentioned earlier, the optical system of previous Mariners is not suitable; long focal lengths are needed to obtain images of sufficient resolution as the spacecraft approaches and recedes from the planet.

Figure 2.5   Fall, 1971. The 44 kilobit capability is dropped, leaving 22 kilobits as the only "guaranteed" rate for the encounter with Mercury and consequently a very poor photo coverage.

The optical system for earlier Mariners had been designed to photograph Mars, a planet with quite different characteristics from Mercury. And because imaging at Venus is to concentrate on acquiring ultraviolet pictures, an ultraviolet capability has to be added. New optics are designed having a focal length of 59 inches (150 cm) compared with about 20 inches (51 cm) for the earlier Mariner system. The longer focal length provides a magnifying effect to show finer detail. The same kind of vidicon tube is used, however, as in the earlier TV system because it has the appropriate image storage qualities to match the data transmission rates of the communications link. This vidicon produces electrical signals to form the television image from the light patterns cast upon it by the optical system of the telescope.

Fall, 1971   The spacecraft manager, John Casani, decides to drop the 44-kilobit Mercury backup rate from the communications system, leaving only two rates: 22 and 117 kilobits. Although the 117 kilobit transmission rate is easy from Venus, it remains only a hoped-for goal from Mercury. Chances of getting good photo coverage of Mercury look poor (Fig. 2.5) without at least the 44 kilobit capacity.

Vic Clarke proposes that the spacecraft should carry a special format on the 117 kilobit link to lower the radio interference threshold. His suggestion is rejected as a source of new complexity and therefore cost.

Imaging team members make a presentation to Dr. William Pickering, Director of JPL, that shows dramatically the advantages of picture quality and coverage that 117 kilobits from Mercury would give (Fig. 2.2).

23

At this meeting, the division manager for telecommunications, Robertson Stevens, is convinced. Puffing thoughtfully at his pipe, he comments wryly:

"The world needs this."

Engineers are confident that they can obtain the needed 117 kilobits through use of the new X-band frequencies.

In view of new developments in electronics, at last the reservations concerning the reliability of the X-band transmitter are overcome. But now there are doubts about the cost of development. A price tag of $2 million is placed on the new system. Since the project does not have contingency funds for such improvements, the additional money is sought from NASA Headquarters. But this proposal is rejected. NASA, too, is in a serious financial squeeze. Congress is proving a hard budgetary taskmaster as economic difficulties grow in the wake of the abortive $1 trillion expenditure in Southeast Asia. NASA insists that the $98 million limit will remain regardless of how much greater capability for data return can be achieved by a 2 percent increase in total cost.

With the X-band capability denied the project, Vic Clarke next appeals to Boeing to increase the size of the high-gain antenna of the spacecraft to 58 inches (147 cm) as on the Viking spacecraft. This is another route by which a higher data rate can be approached. As a performance safeguard, Boeing had already increased the diameter to 54 inches (137 cm). They refuse to make it bigger. No more changes!

And there is still only one flight spacecraft for this mission. In testimony before the House Space Committee, in March 1972, Murray discusses this problem:

"If we were doing it with a backup scheme where you don't necessarily launch the second vehicle, only if the first one fails, if you were to put some money on that, in flight qualification of the second spacecraft the cost might go up [for the dual mission] to $105 to $110 million. The possibility of a successful mission would go up to 90 percent. . . . To me, as a scientist, I am astonished at these economics. It just doesn't make sense to me. The launch failure of Mariner 8 last summer was an unpleasant reminder of the high risk involved."

Later, in 1972, it is becoming apparent that the contingency funds—money set aside for unexpected difficulties in making the spacecraft—are not being used up. Gene Giberson decides that some of this money should now be used to try to get the 117 kilobit communication rate from Mercury. A 35-watt, S-band transmitter project is approved, and a start is made to develop it as an alternative transmitter to be carried on the spacecraft. The dream of 117 kilobits comes a little closer to reality.

24

December 13, 1972 A major crash program starts to develop the 35-watt transmitter. Design and construction continue through June 1973 to ready the more powerful transmitter for the launch date. A final test is scheduled in June. The transmitter is plugged in. Engineers check and double-check. The power is switched on. The transmitter fails!

"I understand how we got there. I understand the reasons. . . . But when you stand back and look at them, especially after the Mariner 8 failure—which took away the optimism that our Mariner launch problem was solved forever—those odds, in terms of dollars, are strange."
—B. C. Murray

Only a few more months to the launch. There now seems no chance of getting 117 kilobit communication from Mercury. There is a second 35-watt transmitter available, but intensive efforts to locate the trouble that caused the first one to fail are in vain. Because the cause of failure cannot be pinpointed, project management cannot give the go-ahead for the second transmitter to be installed in the spacecraft.

There were other important changes made on the spacecraft during development. One of these was to the propulsion system that provides the thrust to change the trajectory of the spacecraft and refine the aim to achieve precise planetary encounter, an essential for the Venus swingby trajectory. Three times as much propellant will now be carried, to provide a total rocket burning time of 550 seconds. The original propulsion system adapted from earlier Mariners has been redesigned, using the same type of propellant tank as in Pioneer Jupiter. The most important change was to increase the attitude-control nitrogen gas supply by 50 percent. This made the extended mission possible.

All along, Mariner Venus/Mercury has been a program with rigorous cost control. To meet the tight budget only one flight spacecraft, one test spacecraft, and a minimum number of spare parts are to be manufactured by Boeing. Spares to support the flight spacecraft are to be cannibalized from the test spacecraft.

However, in March 1973, with only eight months to go to launch, the project looks as though it will be completed at a lower cost than originally estimated. Giberson and others on the project start examining how these cost savings could be used to increase the probability of the mission being successful. Could the test spacecraft be upgraded into a backup flight spacecraft?

"In addition, even aside from the simple business economics, I argue that in terms of our posture in space, in terms of what we have to offer, both historically in terms of time, and also internationally in space . . . this is one of our most promising missions, and yet it's the one we're attaching the highest risk to. I feel that this is a serious omission, and all I can do is express my concern, which I am doing."—B. C. Murray

25

Boeing produces estimates of the cost to upgrade the test spacecraft to flight status. Project management gives the go-ahead when the launch date is only seven months away. To permit the test spacecraft to survive the space environment, Boeing fabricates a complete set of thermal blankets and improves the thermal protection generally—this protection not being needed for a test spacecraft—and upgrades many of the other subsystems of the spacecraft.

June 21, 1973   Good news. JPL is given the go-ahead by NASA Headquarters to start planning an extended mission—second and third encounters with Mercury.

July 30, 1973   Boeing ships the upgraded test spacecraft directly to the Eastern Test Range, in Florida. A week later the flight spacecraft leaves JPL, where it has been going through solar simulation tests, for the Eastern Test Range.

August, 1973   Problems develop in thrusters on Skylab. Components used in the Skylab thrusters are also used in the thrusters of Mariner Venus/Mercury. Mariner's fittings and seals are to be reworked. The changes are completed only one month before launch date.

With the hardware problems of the spacecraft essentially all solved, the battle of the kilobits continues. Shortly before launch, when the prospects of obtaining full photo-coverage at Mercury still seem bleak, Vic Clarke makes a final attempt to improve the capabilities. He talks with Gerald Levy, manager for low-noise receiver cones for the Deep Space Network (DSN) which is operated for NASA by JPL. The network consists of large antennas concentrated near Madrid, Spain, Canberra, Australia, and Goldstone in California's Mojave Desert. These antennas receive the radio signals from the spacecraft and pass the telemetered information back to JPL. From JPL, commands are sent to the DSN stations for transmittal to the spacecraft. Thus the DSN acts as a network on Earth to complete the communications with the spacecraft as the Earth turns on its axis, and first one antenna system and then another faces the distant Mariner.

Clarke asks Levy if he can find any way to increase the performance of the huge, 210-foot (64-meter) antennas of the Deep Space Network (Fig. 2.6) to extract 117 kilobits of data from the whisper-faint radio signal from the spacecraft at Mercury. These radio signals from spacecraft millions of miles from Earth are almost drowned by radio noise arising from space itself, from the distant stars, and from the electrons in the components of the electronic equipment used to receive the signals. The electronic noise is related to the temperature of the devices that first receive the very weak signal. When these are kept extremely cold by immersion in liquid helium, they can detect very faint signals. The lower the temperature, the more sensitive the equipment becomes.

Figure 2.6 Large antennas of the Deep Space Network are essential links in the communication chain from spacecraft to Earth. They rely upon very cold receiving devices to amplify the faint radio signal from the spacecraft and separate it from the interference of radio "noise." Signals arriving from the direction in which the antenna is pointed are reflected off the big "dish" up to the secondary one (seen supported by trusswork) and from there into the multiple receiving cones where amplification can begin.

Clarke asks Levy if he can lower the big antenna's noise temperature level beyond 15.7°Kelvin (that is, 15.7° above the absolute zero of temperature, or −430° on the Fahrenheit scale). Levy retorts that everything possible is being done.

The 117 kilobits from Mercury seem as far away as ever.

The following morning, Clarke is surprised to hear Levy on the telephone. The engineer has wrestled with the problem overnight and consulted with his maser amplifier experts, Walter Higa and Ervin Wiebe. The maser, the device that first receives the radio waves collected by the big antenna, amplifies them at a specific frequency without adding significant amounts of noise. But it has to be kept very cold by immersion in liquified gas. Now Levy tells Clarke that he thinks he has a solution. He believes he can use a super-cold maser and that it can be fitted to the Goldstone antenna. This device might reduce the noise temperature another 2.2 degrees. Moreover, a "mothballed" existing feed that guides the radio waves between the electronics and the antenna, called an "ultracone" and originally used for the Mariner Mars '69 project, can be equipped with a second super-cold maser and shipped to the Canberra station to upgrade the big antenna there.

Clarke is excited and asks Levy to go ahead.

The noise temperature reduction will not quite do the job, however; the 117 kilobit rate is close but not fully assured. Levy says a change in the way the ground station operates is also required. Normally the ground station can both receive and transmit with one antenna. Now the antenna must only receive in order to achieve the high bit-rate desired.

The Mariner spacecraft is to be commanded by an uplink transmitter that operates through the same antenna at Goldstone as is used to receive the signals. This dual "transmit-receive" activity of the antenna makes it necessary to switch the path of the received signal in the microwave "plumbing" system of pipes and tubes mounted inside the antenna dish. The very high frequency radio waves are carried along pipes rather than in wires to reduce losses. Levy explains that if the path can be shortened, the noise temperature can be reduced even more and the 117 kilobits obtained. But this would mean that the spacecraft can not be commanded through the transmitting uplink at the time of encounter with Mercury. The entire command sequence for the encounter will have to be rewritten. Each command, of course, has to be acted upon by the spacecraft and its instruments at precise times that have all been determined in advance. The command sequence is therefore referred to as being "time critical."

Clarke seizes the chance, but it means a great deal of intensive work to modify the station operating plans because launch is now only a few weeks away. Clarke asks Tony Spear, a telecommunications expert, to work out a new plan. The required command sequence is rewritten. Hope now rises among the imaging team that they will get the full coverage and resolution of Mercury's surface that they have sought for years.

Meanwhile, the two spacecraft go through rigorous tests at the Air Force's Eastern Test Range, in Florida, preparatory to launch. The two spacecraft follow somewhat different paths through the test maze. Spacecraft 73-1, which was originally the test spacecraft, goes through its major testing at the Boeing plant and at the Eastern Test Range. System assembly and testing consists of reassembling the spacecraft and testing to make sure it operates as a whole and together with its ground systems. Then the whole system is thoroughly checked and shown to be ready for launch. Afterward the spacecraft is shipped to the test range. Here it is again disassembled, inspected, cleaned, and reassembled, and then checked for systems readiness, compatibility with the Deep Space Network and with the launch vehicle, and put through a complete rehearsal of the countdown.

Spacecraft 73-2, the flight spacecraft, undergoes a similar series of tests; it is put through a countdown simulation, undergoes environmental tests, and completes its systems operational and functional tests at the Boeing facility (Fig. 2.7). It is then shipped during the summer to JPL for systems readiness tests and rigorous environmental tests in a simulator that duplicates the conditions in space and the intensity of the Sun at Mercury's orbit (Fig. 2.8). Later, at the Eastern Test Range, the flight spacecraft is also pulled apart, inspected, cleaned, and put together again, and then put through compatibility tests and a long series of readiness-for-launch tests. Finally it is transferred to the launch pad and mated with the Atlas/Centaur booster. All these tests are completed on October 30, three days before the scheduled launch.

November 2, 1973   The spacecraft is ready. The ground system is ready. All science experiments are ready. The imaging team members are happy about the improved chances for getting 117 kilobit communication from Mercury. The battle of the kilobits has been a long, hard campaign, but victory is at last in sight. Three changes contribute to that apparent victory: the improved low-temperature maser poised on the 210-foot (64 meter) antenna in the Mojave Desert; the reprogramming of the command sequence accomplished at the Laboratory in Pasadena; and a new appraisal of all the communications links and their capabilities made possible by a calibration technique just developed by Vince Evanchuk of JPL. He demonstrated to everyone's satisfaction that there would be less than 3 errors per 100 bits of data received at the 117 kilobit rate from Mercury. These errors arise because the signal will be so faint

Figure 2.7   Every effort has to be exerted to make sure that nothing will fail on the one spacecraft being sent to Mercury. The flight spacecraft is rigorously tested before it is shipped from the Boeing facility. White-suited Boeing test engineers maintain extreme cleanliness of the spacecraft environment throughout the extensive period of testing.

compared to the noise. Three errors in 100 bits had been established by the imaging team years earlier as an acceptable level.

It seems that the stage has indeed been set for the curtain to rise on Mercury. The performers are all ready and in their appointed places, but lurking in the back of everyone's mind is the launch syndrome. Since the earliest space launches there has been a morbid fascination with the thought that something might go wrong and the whole effort might blossom into a ball of flames on the launch pad or shortly after the fire-in-the-tail lifts the big bird into the Florida sky.

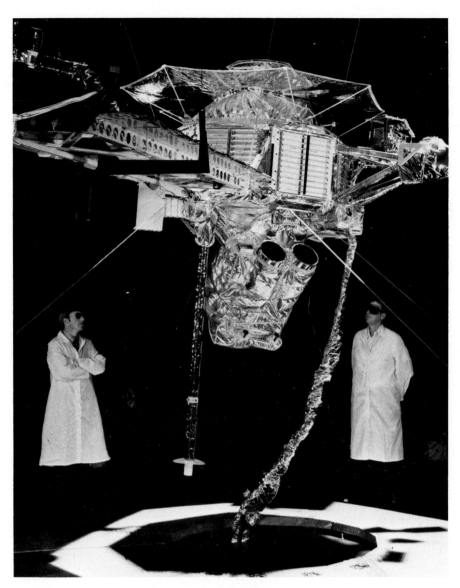

Figure 2.8 Tests continue at JPL where the spacecraft is subjected to the simulated conditions of space, including the intense radiation of 4.8 suns (expected at the orbit of Mercury) on the "top" of the spacecraft and the cold of liquid-nitrogen–cooled walls all around.

Except for this fear, the scientists, engineers, technicians, and managers gathered at the launch site see Mariner Venus/Mercury as being ready for lift-off on its voyage of discovery.

In the seemingly interminable countdown to the launch instant, people's minds stray back to relive many of the events that enlivened that summer of 1973.

3

"We've Got Problems"

Technological advances have made failures infrequent, whether at launch or during the mission. . . . Planetary exploration is no longer a primitive and risky act.—Space Science Board, 1968

October 6, 1973   FOURTH ARAB-ISRAELI WAR BREAKS OUT

October 10, 1973   SPIRO AGNEW RESIGNS AS VICE PRESIDENT OF THE UNITED STATES AFTER PLEADING NO CONTEST TO CRIMINAL CHARGES

October 19, 1973   SAUDI ARABIA CUTS OIL EXPORTS 10 PERCENT TO U.S. AND THREATENS TOTAL EMBARGO IF U.S. CONTINUES TO SUPPORT ISRAEL

October 20, 1973   ATTORNEY GENERAL ELLIOT RICHARDSON RESIGNS. HIS DEPUTY WILLIAM D. RUCKELSHAUS AND WATERGATE SPECIAL PROSECUTOR ARCHIBALD COX ARE FIRED BY PRESIDENT NIXON OVER THEIR ATTEMPTS TO PERMIT WATERGATE TAPES TO BE TURNED OVER TO JUDGE SIRICA

October 20, 1973   LIBYA CUTS OFF ALL OIL EXPORTS TO U.S.

October 22, 1973   KUWAIT, QUATAR, AND BAHRAIN CUT OFF ALL U.S. OIL EXPORTS

October 24, 1973   ISRAEL AND EGYPT AGREE TO SECOND CEASE-FIRE

October 25, 1973   IN RESPONSE TO "AMBIGUOUS" SOVIET ACTIVITIES RELATED TO THE MIDDLE EAST, U.S. MILITARY FORCES ARE PLACED ON TACTICAL ALERT FOR FIRST TIME SINCE CUBAN MISSILE CRISIS IN 1962

October 31, 1973	WHITE HOUSE LAWYER J. FRED BUZHARDT INFORMS JUDGE SIRICA THAT PRESIDENT NIXON CANNOT DELIVER TWO OF THE MOST CRITICAL WATERGATE TAPES BECAUSE "THEY DO NOT EXIST"
November 1, 1973	MANDATORY FUEL ALLOCATION IS IMPOSED IN U.S.; AIRLINES DRASTICALLY SLASH DOMESTIC FLIGHTS
November 3, 1973	EARTH, MERCURY, AND VENUS ORBIT THE SUN TO THEIR APPOINTED PLACES WHERE MARINER 10 CAN BEST BE LAUNCHED FOR ITS DUAL-PLANET MISSION. AS THE SECONDS TICK OFF, THE EARTH'S SPIN ABOUT ITS AXIS ROTATES THE LAUNCH SITE AT CAPE CANAVERAL, FLORIDA, INTO THE UNIQUE POSITION FOR THE LAUNCH: THE THREE INNER PLANETS OF THE SOLAR SYSTEM ARE READY. EVEN EARTH'S MOON IS MOVING INTO POSITION FOR OPTIMUM VIEWING FROM THE INTENDED TRAJECTORY FOR MARINER 10

The hot and humid Florida summer days drag on into November, seemingly unwilling to acknowledge the onset of Fall. Perspiring crews labor at Cape Canaveral to ready the launch facilities and launch vehicle. On the West Coast, engineers and scientists push aside vacation plans—and family needs—to get the Mariner spacecraft and its battery of scientific instruments to the Cape on time. The big Atlas/Centaur rocket, an aggregation of stainless steel balloons balanced on top of sophisticated liquid-propellant rocket engines packing almost 3½ million horsepower, has arrived at Complex 36's pad B in July. The first stage, the Atlas, was originally constructed as a ballistic missile to propel nuclear death over intercontinental distances. But now that solid-propellant Minuteman missiles, buried in silos, have taken over the nuclear deterrent task, this reconditioned military veteran is ready to take new paths through unexplored space in search of knowledge to uplift the spirit of mankind. Will Mercury reveal a record of the ancient history of the inner solar system? Will Venus disclose the secrets of its multiple veils of cloud cover and provide new clues to why this planet evolved so differently from its twin—Earth?

о   о   о

On a rainy January day in 1971 on the University of California campus at La Jolla, Nobel Laureate Hannes Alfven is asked to comment on mankind's continued search for explanations for the origin of the solar system.

"Every civilization of note has devoted a major part of its resources to this task. It is our moral and intellectual duty to retell this story [of Genesis] in modern terms."

о   о   о

34

After Russia's launch of the first artificial satellite, Sputnik I, in 1957, United States space planners realized that a more powerful upper stage would be needed to supplement the payload-carrying capacity of the then secret Atlas ICBM. Convair, who manufactured the Atlas, proposed a new high-performance liquid oxygen/liquid hydrogen upper stage to be named Centaur. In 1959, the Air Force transferred the Centaur program to the newly formed agency for civilian space research and development—NASA. Two years later the first Centaur stood ready to launch.

May 18, 1961
Cape Canaveral
Fifty-five seconds after lift-off from the launch pad, the first Centaur explodes into a fireball.

Development continued. Centaur gradually fulfilled the promise of its name, becoming the high-performance workhorse for the unmanned civilian space effort—it was used to launch Surveyors to soft-land on the Moon, orbiting astronomical observatories, applications technology satellites, communications satellites, and Mariner spacecraft to Mars. Later still it would be used to launch Pioneers 10 and 11 to Jupiter and Saturn, spacecraft that became the first man-made objects to be placed on a trajectory leading right out of the solar system . . . but there were still unpredictable failures for the Centaur along the way.

November 30, 1970
An Atlas/Centaur is launched from Cape Kennedy (originally and now again called Cape Canaveral) carrying NASA's third astronomical space telescope. A nose fairing, called a shroud by space engineers, encloses the highly sophisticated spacecraft to shield it from the supersonic wind velocities created as the big booster accelerates through the atmosphere into space. Astronomers hope this spacecraft will provide further glimpses of the universe at hitherto unobserved wavelengths, rather like a partially deaf man who has heard a few notes of a symphony and is given the ability to hear more. Out beyond the absorption of Earth's atmosphere, the orbiting astronomical observatory will be able to see planets, stars, and nebulas in ultraviolet, infrared, gamma, and X-rays as well as visible light.

One hour later the complicated, sophisticated, expensive observatory burns up in the Earth's atmosphere like a fiery meteor, somewhere over Africa or the Indian Ocean. The reason: the Atlas/Centaur nose fairing failed to separate after penetrating the Earth's dense atmosphere and prevented the spacecraft from being accelerated to orbital velocity. Instead, the spacecraft plunged back with its booster into the atmosphere and met a flaming end. A similar fate had befallen the first American Mars probe in 1964, Mariner 3. Remarkably, the reason for Mariner's shroud failure was diagnosed and a new one was fabricated in time for the companion spacecraft, Mariner 4, to be sent up within the same launch window (the period of time during which a spacecraft can be launched to a target planet for given relative positions of Earth and the target planet).

○　　　　　　○　　　　　　○

June, 1973　Deep in the Mojave Desert of California, electronics engineers and technicians endure blistering heat; the tracking stations were located there to avoid the radio interference of modern civilization. Hurriedly they wrestle with last-minute bugs in the complex circuits hooked to the big antennas at the Goldstone Tracking Station of the Deep Space Network. New electronics have been developed to extract the utmost detail from the very faint radio signals expected from Mariner 10 when it encounters Mercury. Other crews in Australia and Spain toil similarly over their electronics modules. Again and again they rehearse the events expected on the day of launch, checking and double-checking how the equipment will handle the first signals to be received from the spacecraft when it graduates into space from the Eastern Test Range where it will be launched. Everything that might cause problems is corrected. Almost-true-to-life launches are simulated on powerful computers at the Jet Propulsion Laboratory, Pasadena, while the tracking stations follow and collect information from computer-imagined spacecraft lifting from the pad at the Cape and flying into the inner solar system—each imagined mission a computer programmer's science fiction novel written in millions of binary numbers.

At the Boeing plant near Seattle, Washington, it is a beautiful summer day. Snow still glistens in the sunlight on distant Mount Rainier. The teams at Boeing have met the schedule. Their mood reflects the sparkling summer atmosphere of the Pacific Northwest. A short time before, they had not been so happy: when the Mariner spacecraft was being carefully prepared for a deployment test, a knot in a cord abruptly loosened and the magnetometer boom of the spacecraft dropped and hit the ground. Engineers clustered around it, horrified. Thorough inspection revealed that the damage was limited to the boom. There is a hurried replacement so that Mariner can still be shipped, and the spacecraft makes it on time to begin tests in the solar simulator at the Jet Propulsion Laboratory where intensely powerful xenon lamps flood the chamber in man's imitation of the searing rays of the Sun in space. Despite the human efforts to attain perfection, engineers and scientists remember earlier spacecraft, and there is that lingering question . . . is the spacecraft *really* all right?

○　　　　　　○　　　　　　○

July 31, 1969　Mariner 7 is approaching Mars a week behind its companion flyby spacecraft, Mariner 6. Suddenly, silence falls over Mariner 7. The silence continues with occasional interruptions of unusable telemetry for more than seven hours. The spacecraft has lost attitude control and the

36

telemetry is sporadic. Crews at JPL go into a midnight brainstorming session to find out what happened. When attitude control is re-established, 15 out of 90 channels of telemetry are inexplicably lost; the data are just not getting through to Earth. Apparently a battery exploded, damaging the spacecraft with a shower of shrapnel. Venting gas continually disturbs the attitude control system. Even so, ingenuity and hard work permit the flyby of Mars on August 4 to achieve good scientific results. But a complete spacecraft failure had very nearly occurred.

Despite every precaution and a great deal of experience, spacecraft still do unpredictable things in space. Why did the battery malfunction? The cause is unknown. Possibly it was a micrometeoroid puncture, but more likely it was an internal pressure increase in the battery because of a flaw inside it; a flaw that remained undetected through all the tests made before launch.

o   o   o

August 8, 1973 Packed in sections like a precious cargo of paintings, the Mariner 10 spacecraft leaves the Jet Propulsion Laboratory amidst heavy smog, in a convoy of specially equipped vans. It has passed all the final tests. Now it starts its long journey, first to the launch site in Florida, 3,600 miles (5,700 km) away, and then on to Mercury, almost 50,000 times as far. From California to Florida, Mariner 10 averages 50 miles (80 km) per hour. From Earth to Mercury it will average 56,000 miles (90,000 km) per hour.

Once at the Cape, Mariner 10 is coddled and tested like a premature baby, before release into its real world of space. On September 25, technicians transfer it to the Explosive Safe Facility at the launch site. Here the spherical propellant tank of the spacecraft is filled with mono-propellant hydrazine, whose component atoms of hydrogen and nitrogen need but a small amount of catalyst to fly apart and release the tremendous energy of their chemical bonds as heat, forming the rocket jet that will be used to maneuver the spacecraft in flight. Scrupulous cleanliness is maintained to ensure that any contaminants which could act as catalysts are kept out of contact with the propellant. Otherwise the spacecraft could be seriously damaged, even destroyed, by premature explosion of the hydrazine. The interior of the tank and all the fuel lines are cleansed in advance to the sterility of an operating theater.

Release devices are next placed at strategic places on the spacecraft. After launch, squibs (pyrotechnic explosives) will be fired electrically to actuate these devices and release the spacecraft's antenna, instrument boom, and solar panels, and to separate the spacecraft from the Centaur. Finally prepared for space, Mariner 10 is placed within the protective nose fairing—hopefully not a shroud. Again the rigorous tests, this time to make sure that all electrical connections are perfect and that

Mariner is electrically ready for final launch preparations. Instruments within the spacecraft are commanded into testing; the computer and the internal programmer are exercised.

October 23, 1973   The encapsulated Mariner (Fig. 3.1) is transported to the gantry and mated with the waiting Atlas/Centaur. Ten days of intensive testing remain, and they must culminate in the spacecraft's being cleared for launch. Everyone participates—at the launch site, on the pad, and in the blockhouse and control center at Mission Control at the Jet Propulsion Laboratory, at the tracking stations—all manner of people working in close coordination with the launch vehicle, the spacecraft, computers, communications links, big antenna systems. World-wide, people and machines blend together more intimately than a cyborg of science fiction. Will Mariner 10 survive its birth and later thrive in its mission of exploration? Earlier spacecraft had similarly passed successfully through their periods of gestation but succumbed to their birth pangs.

o   o   o

May 8, 1971
9:11 P.M. EDT
Cape Kennedy   An earlier Atlas/Centaur lifts magnificently from launch pad 36B carrying Mariner 8, the first of two American spacecraft intended to orbit Mars and provide the first complete photographic survey of that enigmatic planet. The Atlas performs with customary excellence. So does the Centaur—for a very short time. Beginning with the start of the Centaur's twin engines, about 265 seconds after lift-off, anomalies come thick and fast. The Centaur stage begins to oscillate violently and then tumbles end-over-end. The engines are shut down after only 100 seconds of the planned 543-second burn. The complex spacecraft never fulfills its role as a "space being" but plunges into the Atlantic Ocean about 350 miles (560 km) north of Puerto Rico.

o   o   o

Would Mariner 10 be strangled in its shroud like an earlier Orbiting Astronomical Observatory and like Mariner 3, or be buried at sea like Mariner 8 because of a fickle Centaur—launched from this same pad 36B—or suffer catastrophic failure in flight, like Mariner 7? Earlier failures had been partially circumvented because there had been two Mariners for each mission to Mars. Success was achieved despite the risks. But budget limitations dictate only one planned chance to go to Mercury. The backup spacecraft would not be launched unless any failure of the flight spacecraft could be clearly identified and corrected.

November 2, 1973   Readiness procedures are completed, and destructive explosive devices are installed in the launch vehicle, insurance against failure; insurance, that is, for Florida and for Caribbean Island communities and for ships at sea—certainly not for the mission itself. The dispassionate Range Safety Office stands ready to destroy the launch vehicle in a fiery explo-

38

Figure 3.1 The encapsulated Mariner is transferred to the gantry at the Kennedy Space Center for mating with the Atlas/Centaur launch vehicle.

sion that would engulf Mariner 10, if the Atlas/Centaur should deviate from the planned trajectory so as to endanger shipping, the Florida coastline, or the Caribbean Islands.

November 3, 1973
2:45 P.M. EST

Countdown is picked up at 600 minutes before launch time. Again the series of repetitious tests—but now precisely scheduled on the rigorous countdown clock.

10:30 P.M. EST

Floodlights flash on to pierce the dark Florida night, heavy with humidity. The launch tower gantry rumbles and quivers slightly as the girdered and platformed structure cautiously inches back to reveal the gleaming rocket. Bluish light etches everything against the black sky. Those watching beside the pad are awed by the silent beauty and signifi-

39

cance of this unique twentieth-century cathedral spire presenting an offering to the heavens.

The big rocket points toward Mars, a conspicuous red "star" high in the sky. But it will be sending this spacecraft in the opposite direction, toward the Sun which is now on the far side of Earth.

Two miles away from the launch pad, in the noisy windowless control center, a tense voice gives the command for extremely cold liquefied gases to deluge from high-speed pumps into the cavernous propellant tanks of the launch vehicle. The cryogenic liquids flash momentarily into vapor on encountering what seem by contrast to be red-hot tank walls. Cold gases vent from the rocket and condense moisture from humid air into steam-like plumes.

November 4, 1973
12:44 A.M. EST

Propellant tanks are full. Terminal countdown begins. So far there have been no holds, no stopping of the countdown because of problems—all continues on schedule. Tension is reflected in everyone's faces, scientists, engineers, technicians, controllers, managers. The United States is gambling that one throw of the Mercury-Venus dice would be a winner, that earlier disasters will not be repeated. There is an improvised plan for a backup for Mariner 10, provided only recently. A second spacecraft—an upgraded test version of Mariner 10—is at the Cape, and another Atlas/Centaur, assigned to Intelsat Corporation, has been placed on standby, ready to launch within 36 hours. But these can be used for a second attempt only if the cause for any failure in the launch of Mariner 10 is identified and can be confidently avoided the second time. Since the launch window is open only about one month and often much time is needed to get fool-proof answers (if at all) for space failures, there is really small chance of meeting these rigorous criteria for a second launch if the first should fail. Everyone is fully aware of what is at stake. Orbital configurations of Earth, Venus, and Mercury will not be favorable again for over 10 years. If Mariner 10 fails, mankind will have to wait one or more decades for its first look at Mercury. And for key engineers and scientists, four or five years of their best creative efforts working on the Mariner 10 program will have been virtually wasted.

Launch minus 10 seconds

The launch is now committed. An automatic sequence starts, controlled by computers: lift-off will be attempted without further human aid. In the harsh lights the launch scene looks unreal—like the cover art from *Thrilling Wonder Stories* of the 1930s. All seems quiescent, but the observer senses the presence of immense power waiting to be unleashed. Voices begin to hush in the blockhouse, at the command center, at the tracking stations, in the press centers, at the control center in Pasadena. Everyone waits in a spreading silence broken only by the toneless "five, four, three, two, one," droning over the loudspeakers connected worldwide. Within a second of the time chosen two years before, the triple main engines of Atlas and the two small vernier engines burst into awesome life (Fig. 3.2). Atlas/Centaur number 34 shrugs like a mighty dragon and with agonizing slowness lifts from its pad, lighting up the November night with an unnatural brilliance. These Atlas engines have to burn for several minutes to provide sufficient velocity for the Centaur to take over and accelerate into Earth orbit, where it will momentarily "catch its breath" in a power-off coasting period before sending Mariner 10 on its voyage inward toward the Sun. Two minutes after lift-off the two outer engines of Atlas snuff out on schedule. A minute afterward the main engine also loses its fiery exhaust. Explosive bolts fire on electric command, and the Centaur upper stage separates from the spent Atlas shell. Ten seconds later the brilliant flame of Centaur's twin engines thrusts spears of white light into the night sky, pushing the spacecraft for 9½ minutes until it gains a parking orbit to coast partway round the world at an altitude of 117 miles (188 km). Mariner 10 has survived the birth pangs in which Mariner 8 perished.

1:20 A.M. EST   The Centaur engines erupt into a programmed restart as they are ignited again and thrust the spacecraft from its parking orbit to overcome the gravitational hold of Earth. Two and a quarter minutes of thrust accelerate the spacecraft from 16,480 miles per hour (26,500 km/hr) to the required 25,456 miles per hour (41,000 km/hr). Mariner has now also escaped the fate of the third Orbiting Astronomical Observatory and Mariner 3.

1:28 A.M. EST   Mariner 10 begins to unfold its movable elements—solar panels, antenna, magnetometer boom—like a space creature newly born into its true environment and readying for its life in space. Mariner 10 thus enters its cruise phase. It has been a perfect launch!

"Thank God! We've made it!"

Control of Mariner 10 passes from the Cape to JPL in Pasadena. Engineering information pours back to Earth over the telemetry links. At the Space Flight Operations Facility in Pasadena, project engineers anxiously scan the data displayed on television screens and inspect the mounting piles of folded paper copies which the computer printers chatter out. The spacecraft looks very healthy.

Kennedy Space Center falls quiet and dark as scientists and engineers go back to their motels where many celebrate the successful launch with free-flowing champagne toasts. Only the clean-up crew at the launch pad still must work. Everyone else goes a little crazy now that the weeks and months of pervasive tension are over . . . but the bone-deep accumulated fatigue begins to gain the upper hand.

"At last, some real sleep. Then if only the weather holds so we can enjoy those Florida beaches we've been driving by every day."

41

Back in California, the Mariner 10 activity continues. Mission controllers are sending commands to the spacecraft to turn on some of the scientific instruments and calibrate them in the well-known environment of the Earth-Moon system.

November 4, 1973 *NEW YORK TIMES* CALLS FOR PRESIDENT NIXON'S RESIGNATION

November 4, 1973
4:45 A.M. EST
The spacecraft's star tracker senses the bright northern hemisphere star Vega, after another sensor has locked on to the Sun. Mariner 10 now locks onto Vega and becomes firmly oriented in the no-up, no-down environment of space so that scientific instruments, the high-gain antenna, and the solar panels can all point in the required directions. Later the bright southern star Canopus will be used instead of Vega to orient the spacecraft.

Later, in Florida, Bruce Murray's phone rings:

"Bruce, this is Ed" [*Ed Danielson is TV Experiment Representative at JPL*].

"What time is it, Ed?"

"It's five o'clock Sunday morning here in Pasadena; that's 8:00 a.m. your time. . . . We've got problems. The TV heaters haven't come on."

The two television cameras (Fig. 3.3), heart of the mission's scientific instruments, have aluminum lens barrels. Like a small electric blanket surrounding each barrel is a strap heater, to keep each at the tempera-

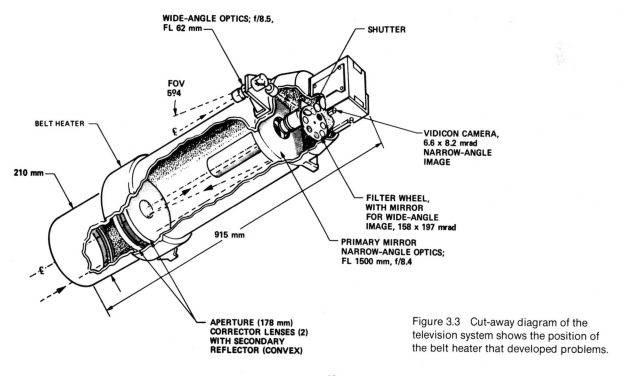

WIDE-ANGLE OPTICS; f/8.5, FL 62 mm

SHUTTER

FOV 5°4

BELT HEATER

210 mm

VIDICON CAMERA, 6.6 x 8.2 mrad NARROW-ANGLE IMAGE

915 mm

FILTER WHEEL, WITH MIRROR FOR WIDE-ANGLE IMAGE, 158 x 197 mrad

PRIMARY MIRROR NARROW-ANGLE OPTICS; FL 1500 mm, f/8.4

APERTURE (178 mm) CORRECTOR LENSES (2) WITH SECONDARY REFLECTOR (CONVEX)

Figure 3.3 Cut-away diagram of the television system shows the position of the belt heater that developed problems.

43

Fig. 3.2 Within a few thousandths of a second of the preplanned time, Mariner 10 is on its way to Mercury.

ture required for effective operation. This heating is essential because the cameras are located in the permanent shade of the spacecraft to protect them both from scattered light and, most importantly, from the searing solar radiation and heat. At Mercury the solar intensity will be five times what it is at Earth's distance from the Sun. Thus, Mariner 10's cameras have been designed to operate in a celestial freezer instead of a solar furnace—there is nothing in-between in space. The heaters are essential to keep the camera systems at the design temperature of 40 to 60°F (4–15°C).

"The heaters aren't working and we don't know why."

Mission controllers scanning the engineering telemetry data from Mariner know that the heaters have not come on as commanded. But engineering data do not give clearcut clues why. Quickly a command is sent to deactivate the heaters, followed by another command to try to switch them on again. Still no electric power to the heaters and no heat reaches the cameras. The temperature sensors on them transmit disturbing information. Danielson relays it to Murray in Florida.

"The temperatures are dropping—down to 15° below zero already on the front of the optics."
"What do the preflight lab tests indicate, Ed?"
"The cameras should go completely out of focus soon and stay that way."

8:00 A.M. PST Engineers from Boeing and the Jet Propulsion Laboratory scrutinize the printouts from the computer. Still no solution is found. The TV science team members, spacecraft experts, and camera design engineers are all baffled as to the reason the heaters are not receiving electricity. Telephone calls to engineers in Seattle and all the ingenuity that can be mustered fail to produce any other way to pipe heat to the ailing cameras. The thermal insulation of the spacecraft is so well designed that there is no way to heat the cameras from other parts of Mariner 10. Meanwhile the spacecraft has started its camera calibration series of pictures of the Earth and the Moon. Murray again is on the telephone to JPL.

"Get some people from IPL [Image Processing Laboratory] processing the Earth and Moon pictures as soon as they are received and see if we can identify defocus effects. Tell Giberson we may be faced with catastrophic failure and that we should know within 48 hours. . . . Tell him the possibility of a back-up launch must be kept open."

The main question is: How cold? And what would be the effect of this chilling? Will the precisely fabricated cameras defocus as their material chills and contracts? Will this heater failure be catastrophic to the mission?

Figure 3.4   Mosaics of Earth were sharply in focus. The heater problem had not degraded the images.

Figure 3.5a   Mosaics of the Moon revealed excellent detail to the sunrise terminator.

Figure 3.5b   Computer-enhanced strip image of 3.5a.

47

"I'll catch the next plane back."

Murray packs and heads for California.

1:58 P.M. PST   The first TV pictures of Earth are taken with the cold cameras and radioed from Mariner 120,000 miles (193,000 km) away. The pictures are displayed on TV screens, then converted into prints. They show the sunlit, cloud-covered Pacific on the horizon, looking as peaceful as ever, with the line of sunset—the evening terminator—crossing the Gulf of Mexico. At JPL Mission Control, Danielson and other weary, sleep-starved members of the imaging team peer through magnifying lenses. The pictures look good. They seem to be in sharp focus (Fig. 3.4). So far the temperature problem has not degraded the camera optics. Perhaps the lens elements and the optical tube elements are compensating oppositely—more than designed—for the temperature changes, thus preserving the focus. But still there is danger. The invar rods that support the optics elements of the telescope system and the potting compound used to protect components of the vidicon of each television camera will deteriorate at low temperature. That danger point is −40°F (−40°C). As the hours pass, the telemetered record shows that the temperature slowly is dropping lower and lower, while engineers and image team scientists try to compute how low it might fall.

Mariner 10 is 70,000 miles (112,000 km) from the Moon (Figs. 3.5, 3.6). Its vidicon captures a view of the Mare Humboltainum in densely cratered northern lunar highlands and the picture is transmitted to Earth. Close inspection reveals superb quality, but with a small, peculiar distortion in the pictures from one of the cameras. The cameras are at the threshold of defocusing. The temperatures in the cameras still fall, and the teams working around the clock at JPL try to determine whether the cameras will survive. In the meantime, the vidicons are being kept operating so that heat from their filaments will help to warm the cameras slightly. It is feared that if the vidicons were allowed to cool by switching them off between picture-taking sequences, as originally planned for the mission, they might crack when turned on again. But will the vidicons have to be kept on all the way to Mercury? Not rated for such continuous operation, they may wear out.

November 5, 1973   The temperature of the cameras stops falling and stabilizes at tolerable
9:45 P.M. PST   levels: −22°F (−30°C) at the front of each telescope, −4°F (−20°C) at the back of the optics, and 14°F (−10°C) at the vidicons. Mariner can make it—if the vidicon tubes do not wear out over the next five months of continuous use!

November 5, 1973   DONALD SEGRETTI SENTENCED TO 6 MONTHS IN PRISON FOR ILLEGAL "DIRTY TRICKS" ON BEHALF OF RICHARD NIXON IN 1972

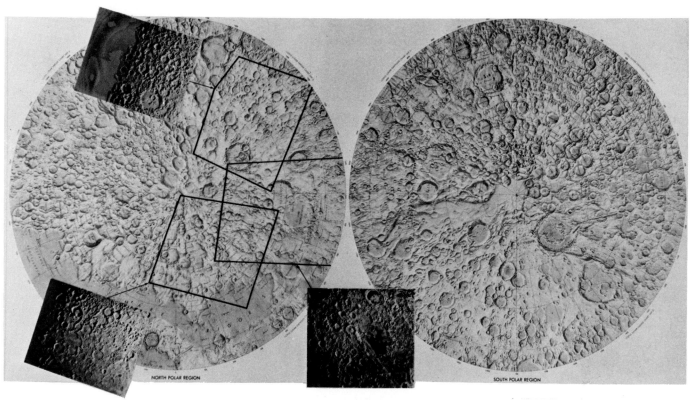

Figure 3.6 Mariner's photos of the Moon's north polar region fill in some of the details that could not be covered by earlier space flights. (NASA lunar polar chart)

NORTH POLAR REGION

SOUTH POLAR REGION

November 7, 1973 8:00 A.M. PST	Mariner 10 is 1 million miles (1,609,000 km) from Earth. It has now sent back 700 TV pictures, together with many ultraviolet scans of the Earth, and has observed electrical and magnetic properties of space during its passage through the Earth-Moon system. Now the spacecraft is commanded to start star photography of the "Seven Sisters" Pleiades cluster in Taurus. An 84-picture sequence displays star images in sharp focus, further confirmation that the "eyes" of Mariner 10 have survived frostbite, further hope that man can indeed get his first close-up glimpses of Mercury and Venus through those hardy television eyes.
November 7, 1973	PRESIDENT NIXON ANNOUNCES "PROJECT INDEPENDENCE" TO ACHIEVE ENERGY SELF-SUFFICIENCY FOR THE U.S. BY 1980
November 13, 1973	The first trajectory correction maneuver, TCM for short, is scheduled for mid-afternoon. When released from the Centaur, Mariner 10 was on a trajectory that would take the spacecraft past Venus at about 30,000 miles (48,300 km) on the sunward side of the planet, arriving closest to Venus about three hours later than the desired time. Neither property of the trajectory would be suitable for the Venus swingby acceleration

49

needed to reach Mercury. Mission controllers patiently waited several days before preparing corrections. During this time, tracking data from the Deep Space Network stations provided measurements to determine the precise orbital path. The TCM would increase Mariner 10's speed by about 24 feet per second (7.3 meters/sec). This change will cause Mariner to arrive at Venus three hours sooner. In addition, the TCM will be directed to change the path of Mariner 10 just enough to send the spacecraft through the desired aim point. The result of the TCM is that the spacecraft's closest approach to Venus will be 6,600 miles (10,560 km) over the dark side of the planet.

Normally the spacecraft is oriented in space by two sensors. A star sensor looks at the bright southern star Canopus and provides a reference line to that star. A sun sensor provides a reference line from the spacecraft to the Sun. The spacecraft is said to be locked onto Canopus and the Sun. But lock is broken if the spacecraft is rotated so that the Sun and Canopus become out of view of the sensors. The TCM requires breaking lock on the Sun and Canopus, and then pitching and rolling the spacecraft so that its rocket engine will direct thrust in the proper direction for the trajectory change.

3:00 P.M. PST A command received by the spacecraft, one of 1,019 sent earlier and stored in Mariner 10's memory for this TCM, starts the gyros so that the spacecraft will know how many degrees to turn through on its pitch and roll axes. Next the telemetry transmissions are disconnected from the high-gain antenna and commanded to pass through the low-gain antenna, since the pitch and roll maneuvers will turn the high-gain antenna beam away from the Earth.

4:08 P.M. PST Nitrogen gas jets spurt momentarily from the tips of the solar panels. The spacecraft begins to roll. When it has turned through 49 degrees, opposing jets spurt their nitrogen gas into space and the roll is stopped. At 4:21 P.M., gas jets spurt from other cold thrusters on the outriggers of the spacecraft, and it now begins to pitch slowly for 12 minutes until it has turned through 127 degrees. Again nitrogen gas pulses into space and stops the pitch. This pitch and roll has pointed the main rocket engine in the desired direction. At 4:42 P.M., a valve opens and allows hydrazine propellant to pour into the reaction chamber. The catalyst there releases the propellant energy. Expanding rapidly, the hot gas rushes from the engine's exhaust nozzle and pushes the spacecraft in the opposite direction. This "burn" lasts 20 seconds, just enough to produce the required velocity change. Then the cold-gas thrusters pulse again and return the spacecraft through pitch and roll "unwinds" until the spinning gyros tell the spacecraft that it has returned to the orientation it started from. Next the spacecraft's sensor has to find Canopus and lock onto it again.

5:08 P.M. PST   The star tracker on Mariner 10 reacquires Canopus. The first TCM is completed; Mariner 10 has passed another key phase in its growth, like a child's first step. Suddenly the data indicate that the spacecraft has thrown a tantrum; it has gone into an uncommanded roll search. The star sensor has lost Canopus! Quick action follows in the control center. But more than an hour and a half pass before the telemetry tells mission controllers that the spacecraft once again has reacquired the bright star. Commands that had been deleted during this disorientation now have to be reestablished in the spacecraft so that the mission time line can be brought back to normal. Subsequent analysis reveals that the star tracker has been tricked by a nearby bright particle, probably a piece of dust shaken loose from the spacecraft during the burn of the rocket engine during the TCM. The star tracker, mistaking the particle for Canopus, tried to align the spacecraft to the moving particle instead of the fixed star.

November 16, 1973   PRESIDENT NIXON SIGNS INTO LAW A BILL GIVING FINAL APPROVAL TO THE CONSTRUCTION OF THE CONTROVERSIAL ALASKAN PIPELINE; COURT REVIEW OF ENVIRONMENTAL IMPACT IS SPECIFICALLY BARRED

November 21, 1973   Mariner 10 still performs admirably except for the inability to supply electrical power to the TV heaters. The ultraviolet spectrometer team is preparing to make some scans of the planet Mars in a few days as a further check on the calibration of their instrument. The navigation team is amassing data on the trajectory and refining the calculations for the second TCM. At about this time the gyros are commanded to turn on to put the spacecraft through one of its routine roll calibration maneuvers. Unexpectedly, one of the spacecraft's major subsystems, called the flight data system, resets itself automatically to minus four—like an electronic calculator suddenly going to minus four in the midst of a calculation. Controllers quickly issue commands to cancel the roll calibration maneuver. The telemetry records are scrutinized, but engineers cannot determine whether this unexpected reset indicates a real problem in the power system of the spacecraft or instead is caused within the flight data unit itself by a too-sensitive sensor that "thought" there was a power problem. But any power problem in a spacecraft intended for a long mission can easily be disastrous. Project management decides to play safe and abandon the roll calibration maneuver so that more engineering data can be collected from the spacecraft for several days with the hope of showing what the problem might really be. But no explanation is found—another mysterious problem. Can there be something basically wrong with this spacecraft?

November 21, 1973   WHITE HOUSE DISCLOSES AN UNEXPLAINED 18½ MINUTE GAP DISCOVERED IN A KEY WATERGATE TAPE OF A NIXON-HALDEMAN CONVERSATION

November 26, 1973	STOCK MARKET DROPS TO LOWEST POINT IN 11 YEARS
December 6, 1973	GERALD R. FORD SWORN IN AS NEW VICE-PRESIDENT OF THE UNITED STATES
December 14, 1973	Mariner's solar panels are commanded to tilt by about 18 degrees away from the Sun to prevent their becoming too hot—another of Mariner 10's many adaptations to penetrate the innermost regions of the solar system.
December 19, 1973	AFTER 15 YEARS AS GOVERNOR OF NEW YORK, NELSON ROCKEFELLER RESIGNS TO PURSUE "A NEW NATIONAL ROLE"
December 19, 1973	Mariner is commanded to perform its fourth roll calibration maneuver. Strangely, there is no automatic reset of the flight data system, as occurred uncommanded in two similar previous maneuvers. But no explanation can be found.
December 25, 1973	ARABS LIFT OIL BOYCOTT, EXCEPT AGAINST THE U.S. AND THE NETHERLANDS, BUT CRUDE OIL PRICES ARE DOUBLED

"Bruce, this is Ed. I'm at JPL."
"On Christmas Day? Don't you ever go home?"
"The antenna power just dropped 6 db [to one-quarter]."
"Sometimes I'm not really sure God is on our side in this mission."

Shortly before 1:00 P.M. PST, something had happened aboard Mariner that dropped the signal strength radiated from the high-gain antenna. Hurriedly called to Mission Control, engineers make tests and conclude that a joint in one of the feed systems carrying the radio waves from the spacecraft's transmitter to its antenna may have cracked or fractured. Probable cause: temperature differences between sunny and shaded surfaces.

"There goes our real-time picture sequence at Mercury."

Although this failure would not be too serious at Venus, it would cause serious problems at Mercury. At the great distance of Mariner 10's encounter with the innermost planet every watt of signal from the spacecraft will be needed to transmit the high-resolution pictures of the planet's surface while the spacecraft is actually flying past it, in real time. With the feed system awry, the required 117 kilobit transmission rate is not possible. As much as 80 percent of the picture data at Mercury may now be lost. It seems that the imaging team has lost the "battle of the bits" after all. And an even worse doubt begins to gnaw at everyone:

"What's to keep the antenna from quitting altogether so that Mariner 10 will fly by both planets—silently?"

December 29, 1973	Abruptly the radio signal from the spacecraft comes back at full signal

strength. The antenna problem has healed itself! But four hours later, as abruptly as it healed, the antenna fails again.

January 3, 1974   Mariner 10 is 13 million miles (21 million km) from Earth, 18 million miles (29 million km) from Venus, and 61 days into the mission. At 5:50 P.M. PST, the high-gain antenna problem suddenly disappears.

January 6, 1974
11:33 P.M. PDT   Again the high-gain antenna fails despite continued attempts to keep it warm by solar heating. The specter of greatly reduced picture coverage of Mercury once more hovers ominously. Quickly, alternate sequences are developed for the Mercury encounter. One of them assumes a rate of only 7.35 kilobits per second at Mercury encounter.

January 8, 1974   PRESIDENT NIXON ISSUES TWO PAPERS DENYING ANY WRONGDOING IN THE MILK FUND AND ITT CASES

January 8, 1974
7:39 A.M. PDT   Mariner 10 unexpectedly switches automatically from its primary power processing system to its redundant energy backup system. Reason: unknown. Now the spacecraft must rely upon that single power supply, upon only one "lung." If that fails also, then the mission will be terminated—the "space being" will die. No such failure has been encountered previously in the Mariner program.

January 9, 1974   Mariner 10 begins its observations of Comet Kohoutek (Fig. 3.7) with passive measurements of ultraviolet radiation from the comet's tail.

Most comets are discovered only a few months ahead of their closest approach to the Sun. Comet Kohoutek is unusual in that it was first seen an unprecedented nine and a half months before it would make its perihelion passage. Because it was seen so far out on its orbit, astronomers speculated that it was very large, perhaps the largest comet in nearly a century.

NASA has four concurrent space activities that can observe Kohoutek: Skylab, the manned orbiting space station; Copernicus, the successor to the ill-fated Orbiting Astronomical Observatory 3; Orbiting Solar Observatory 7; and Mariner 10. All could observe the comet in regions of the spectrum denied to ground-based observers. As Comet Kohoutek comes closer, it proves to be almost 40 times dimmer than expected, removing any chance of photography from Mariner 10. Nevertheless, it is the best-studied comet in history.

The prime objective of the Mariner 10 observations of Kohoutek is to gather unique data in the extreme ultraviolet region of the spectrum which cannot be obtained from Earth because of absorption in Earth's atmosphere. Moreover, as opposed to Skylab, Mariner 10 is well outside the Earth's own hydrogen corona (which masks the hydrogen in the comet), thereby enhancing further the spacecraft's capabilities to observe Kohoutek.

Figure 3.7 Comet Kohoutek was investigated by Mariner to show details of ultraviolet radiation from the comet.

January 15, 1974   SIX COURT-APPOINTED EXPERTS REPORT THAT THE 18½ MINUTE GAP IN THE NIXON-HALDEMAN TAPE COULD NOT HAVE BEEN ACCIDENTALLY PRODUCED

January 22, 1974   Mariner 10 begins to be readied for its upcoming Venus encounter. The movable scan platform on which the TV cameras are mounted is given its final pointing calibration and the cameras take three sequences of test pictures of star clusters.

January 28, 1974   Mariner 10 is commanded to conduct its final roll calibration maneuver before its Venus flyby. Eight rolls are performed in 79 minutes. At the end of each roll the scan platform is moved in a cone to obtain data on the diffuse emissions of ultraviolet light over wide regions of the sky. Suddenly, a computer alarm alerts engineers to a rapid decrease in maneuvering-gas pressure aboard Mariner 10. The roll-gyro rate telemetry channel shows violent oscillations, as a result of which the irreplaceable nitrogen gas supply has dropped from 6.0 to 4.7 pounds (2.7 to 2.1 kg). The loss stopped when the gyros were turned off. What now? Will this unprecedented problem affect the capability of the spacecraft to make further trajectory correction maneuvers? Two such course corrections are essential between Venus and Mercury.

After almost a week of nearly continuous analysis of the telemetered data and the characteristics of the spacecraft's systems, project management decides that the gyros might be operated without exhausting the attitude control gas. But the attitude control problem and the attendant risk of losing all the precious nitrogen gas if celestial reference should ever be lost causes the controllers to develop a number of changes and "workarounds" in mission operations. Further roll calibration maneuvers are canceled.

With the Venus encounter approaching, an important decision now has to be made: should Mariner 10 fly past Venus controlled by the gyros as originally intended, or, in view of the roll oscillation problem, should the flyby be attempted with the spacecraft locked on to the Sun and Canopus for stabilization? If the gyros are used, the gas-consuming oscillations might recur and not only deplete the remaining reserves and prevent a subsequent encounter with Mercury but also introduce unwanted oscillations of the spacecraft that would spoil much of the flyby observation of Venus. If the Canopus sensor is used instead, it could well be distracted by the brightness of Venus and cause the spacecraft to turn toward the planet instead of the star. This also would have disastrous consequences—the need to use the gyro system to overcome the Venus distraction and try to retrieve lock on the star, and loss of some, perhaps most, of the data to be gathered during the Venus encounter.

February 1, 1974   Mariner 10 is 1,690,000 miles (2,719,000 km) from Venus and is approaching it at 20,000 miles per hour (32,000 km/hr), constantly picking up speed. Project management has decided that the Venus encounter sequence on February 5 will be performed with the spacecraft oriented on the star Canopus and the Sun; no gyros will be used. Revised commands for this improvised flyby sequence are being transmitted to Mariner 10's electronic memory.

Venus flyby is now scheduled to take place Tuesday, February 5, 1974, with closest approach at 10:01 A.M. PDT. Venus appears dark to Mariner 10 as the spacecraft speeds toward the planet from the night side. The true test of spacecraft stability will come as Mariner 10 rushes toward the daylight side of Venus and the crescent of Venus starts to form, bathing the spacecraft in brilliant reflected sunlight.

Meanwhile, everything else is fine—except for the TV heaters—the fact that the antenna power is down—and that one electrical power system is dead—and the platform on which the TV cameras are mounted cannot be raised as much as we want—and— Doubts proliferate. Will we really get a first close-up look at Venus, much less Mercury?

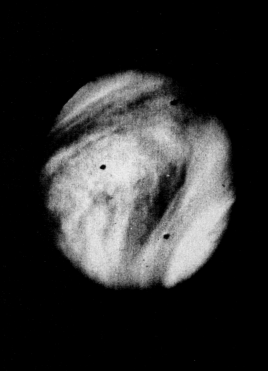

4

Color Venus Ultraviolet

Little is known about the structure of the Venus clouds. Uncounted visual observations and thousands of photographs of Venus in the visible have produced only one frustrating result; a yellowish, featureless cloud deck. In the invisible ultraviolet, however, indistinct markings can sometimes be photographed.
—Mariner 10 Imaging Team, 1971

In 1790, shortly after the United States of America first stepped out on the stage of history, an obscure European, Johann Schröter, discovered that the thin crescent of Venus, when appearing near to the Sun and approaching inferior conjunction (when the planet is between Earth and the Sun), could be seen to extend far beyond a half-circle. He correctly surmised that this was a result of a Venusian atmosphere illuminated by sunlight refracted and diffused from the sunlit side of the planet that was facing away from Earth. Later he saw this "twilight" glow span the dark limb, or edge, of Venus in a complete arc. Even earlier, in 1761, the Russian Mikhail V. Lomonosov suggested (from observations of a transit of Venus across the face of the Sun) that Venus has an atmosphere. And in the seventeenth century a Dutch scientist, Christian Huygens, had said that the unblemished appearance of Venus indicated an atmosphere surrounding the planet.

In the following centuries little more could be inferred about this Venusian atmosphere except that it seemed remarkably thick and impermeable. Visual observations over all these years, coupled with thousands of photographs in visible light in the present century have shown only a yellowish, featureless covering of clouds (Fig. 4.1).

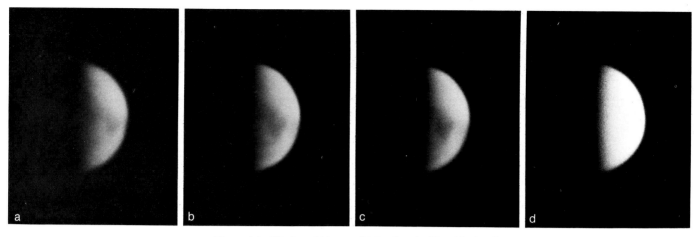

Figure 4.1 Photos of Venus obtained at the New Mexico State University Observatory on the night of 7–8 June, 1967, reveal the faint atmospheric markings and relatively rapid rotation seen in ultraviolet light at 2230 on the 24-hour clock (4.1a), 0206 (4.1b), and 0311 (4.1c). Figure 4.1d, taken in green light at 0150, shows a surface completely devoid of features.

Surprisingly, Venus does exhibit features in wavelengths beyond the blue limit of the human eye, in ultraviolet. As long ago as 1913, F. Quenisset at the Flammarion Observatory, Juvisy, France, first photographed such faint and diffuse markings on Venus. In 1935 Earl Slipher, at Lowell Observatory, found rapid changes of features in ultraviolet photographs over a period of a few hours. As further ultraviolet photographs were amassed over the years, some astronomers claimed that there was a distinct pattern suggesting a rapid rotation of these markings in a period of several days. By 1970, C. Boyer and P. Guerin concluded that the ultraviolet markings were not only real but persistent. Moreover, these markings rotated around Venus in a period of 4 or 5 days in the retrograde direction (i.e., clockwise, counter to the direction of revolution of all the planets around the Sun and to the direction of rotation of the Sun and nearly all the planets on their axes). Other observers, however, could not identify such a clearly defined pattern in their ultraviolet photographs. They suggested that the ultraviolet markings might not be related to the motion of the main atmosphere but rather might arise from unrelated stratospheric phenomena like noctilucent clouds on Earth, high clouds that glow faintly in the night sky.

o   o   o

On May 10, 1961, a large antenna at the Goldstone Tracking Station in the Mojave Desert beamed a radar signal at the planet Venus. A few minutes later the antenna received a faint echo. For the first time in history a radar echo had been detected from another planet. Mathematical analysis of the returned signals' characteristics indicated that Venus rotates at an extremely slow rate, perhaps even synchronously with its revolution around the Sun, in the same way Mercury was then believed to rotate.

Other radar astronomers world-wide repeated the experiment.

> Experimental results for the period November 20 to December 12, 1962, indicate that if the axis of rotation of Venus is perpendicular to the plane of the ecliptic, there is most probably a reverse rotation [rotation in a direction opposite to the motion of Venus around the Sun] with a period of 200–300 days. (V. A. Kotel'nikov, et al., *Reports of USSR Academy of Science*, vol. 151, no. 3, 1973).

When Venus was near the Earth in 1962, the Goldstone antenna was again used to send and receive signals from the planet; R. L. Carpenter and R. M. Goldstein were able to prove that the rotation of Venus is retrograde with a period of about 243 days. Later observations by Goldstein and others have established the rate as 243.1 days, extremely close to the synodic period of Earth and Venus (that period between the successive appearances of Venus in the skies of Earth)—Venus always seems to face the same hemisphere toward Earth at the time of closest approach of the two planets.

The slow planetary rotation derived from the radar results made it very difficult to understand (and therefore to accept) a rapid and organized rotation of the upper atmosphere as suggested by the ultraviolet markings. Unlike Earth and Mars, the winds of Venus appeared to lead rather than follow the planetary rotation. It is as though a weather map on Earth were to repeat periodically and to rotate some 60 times *faster* than the surface of the planet.

Meanwhile, the interplanetary age was beginning. On February 12, 1961, the Soviet Union launched Venera 1 toward Venus. The probe was lost a few days later when Soviet ground stations could not maintain radio contact.

On July 22, 1962, the first American spacecraft for Venus, Mariner 1, was launched from Cape Canaveral. Unfortunately, the launch vehicle strayed from its path. Precisely 293 seconds after lift-off the range safety officer pressed a button and the Atlas/Agena blossomed into a ball of flame. On December 14, 1962, a second American attempt, Mariner 2, passed within 21,598 miles (34,557 km) of Venus to become mankind's first probe of another planet. No appreciable magnetic field was found despite the close similarity of Venus to Earth in size and mass (and probably in internal constitution). The evidence from Venus suggested that a rapid planetary rotation rate, like Earth's, is necessary to produce the self-sustaining internal dynamo envisioned to explain Earth's magnetism. But then, on July 15, 1965, Mariner 4 (the first probe to Mars) found that it too lacked a magnetic field even though Mars's rotation rate is nearly identical to Earth's. Apparently both an Earth-like density *and* a rapid rotation rate are required. Mercury, with a small mass like Mars and a slow rotation like Venus, was deemed therefore a very unpromising prospect to exhibit an Earth-like magnetic field.

Finally, after 11 frustrating failures of spacecraft aimed for Venus, the Soviet spacecraft Venera 4 reached its destination on October 18, 1967, and successfully released an entry capsule into that planet's mysterious atmosphere. The Soviets announced that the capsule reached the surface and measured a pressure 18 times that at Earth's surface and a temperature of 540°F (280°C). Also, Venera 4's capsule found that the atmosphere of Venus is composed almost entirely of carbon dioxide. Formidable as the environment sounded, it was still much more hospitable than many American scientists had expected, based on their analysis of the steady background of radio noise observed from Venus.

Two days later Mariner 5 reached Venus and provided new information. When the spacecraft passed behind the planet as viewed from the Earth, as planned, the radio signal was occulted by the limb of the planet. The dense atmosphere refracted the radio signals to and from the spacecraft, somewhat similarly to the way Earth's atmosphere distorts the visible image of the setting Sun. Measurements of the refraction effects on the radio signal combined with precise tracking of Mariner 5 and with radar-determined measurements of the exact diameter of Venus made it clear that Venera 4 had still been far above the surface when it measured a pressure 18 times that of Earth's atmosphere at sea level, confirming the ground-based radio astronomy interpretations of much higher surface pressures. The real surface was about 16 miles (26 km) farther down, where the pressure is nearly 100 times that at sea level on Earth. The temperature at the surface of Venus is 750–930°F (400–500°C)—hot enough to melt lead and to cause some of the planet's surface materials to glow a dull red in the dark.

For several years afterward, Soviet articles and reports continued to state that Venera 4 had reached the surface of Venus—until Venera 5 and 6 in May 1969, Venera 7 in 1970, and Venera 8 in 1972 all confirmed the much greater pressures and temperatures derived from the Mariner 5 occultation experiment.

○　　　　　　○　　　　　　○

None of the successful Soviet or American Venus probes carried a camera, however, to provide a close visual inspection of the planet. The question of the nature of the ultraviolet markings and whether there were visible cloud features remained unanswered. Actually, Mariner 5 did at one time contain a camera system, when it was the backup space-craft for the 1965 Mars mission carried out by Mariner 4. It would have been a relatively simple matter to modify that small optical system so that it would have been able to obtain a modest number of ultraviolet as well as visible pictures of Venus at the 1967 opportunity. But the scientists advising NASA showed so little enthusiasm for studying a planetary atmosphere by photography that the cameras and accompanying equipment were *removed* from Mariner 5 before it was launched to Venus and were replaced by a second radio occultation system which, as it turned out, did not much enhance the results obtained by the primary one. Close-up photography was not attempted for nearly seven more years, until 1973, when Mariner 10 would try to observe Venus on the way to Mercury.

February 4, 1974   PATRICIA HEARST IS KIDNAPPED BY THE SYMBIONESE LIBERATION ARMY

February 4, 1974   Final update commands are sent to Mariner 10 before its encounter
10:20 A.M. PDT   with Venus.

10:00 P.M. PDT   Television main power is turned on at the spacecraft.

Telemetry from the ailing spacecraft is watched carefully at the Mission Control Center. Mariner 10 hastens toward the night hemisphere of Venus. There is as yet no danger of distracting the Canopus tracker by scattered light from Venus; that test lies 12 hours ahead.

February 5, 1974   Commands are relayed from the spacecraft's memory to set the filter
8:51 A.M. PDT   wheels and exposure time for both cameras. The cameras and the data system are readied to take the first pictures of Venus.

Mariner 10 is rapidly approaching its encounter with Venus from the night side of the planet (Fig. 4.2). As seen from the spacecraft, the dark bulk of the planet is expanding quickly to engulf more and more stars ahead. If an astronaut were aboard, he would see a pearly bright ring of atmosphere outlining the dark globe—Schröter's twilight ring is at close hand nearly two centuries after he first recognized it from Earth. An air of tension now pervades the control center at JPL. Even though mission controllers are not receiving pictures from the spacecraft, they know that the critical time is approaching when that pearly twilight glow will change to a brilliant arc as sunrise on Venus is observed from the spacecraft. Then will be the moment of truth. Will the difficult analysis which indicated a small probability of Venus's brightness pulling the Canopus star tracker off Canopus prove accurate in practice? Will the sensor indeed withstand the brilliance of sunlight reflected from Venus's cloud decks? If so, Mariner 10 will sail past unperturbed. If not, Mariner 10 will fail to obtain near-encounter images of the clouds of Venus, perhaps become disabled permanently.

61

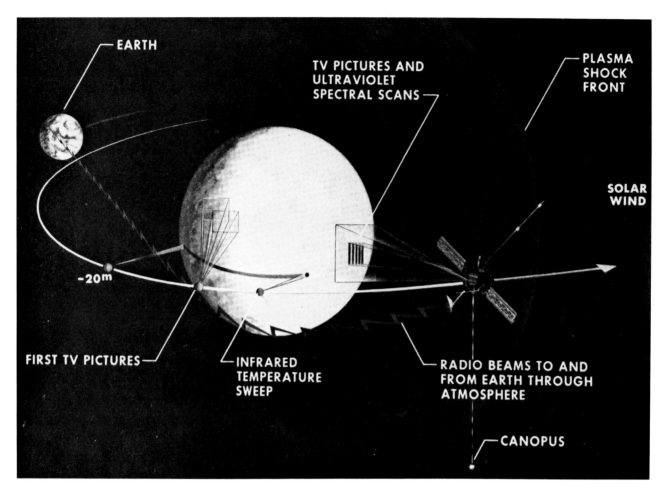

EARTH

TV PICTURES AND
ULTRAVIOLET
SPECTRAL SCANS

PLASMA
SHOCK
FRONT

SOLAR
WIND

−20ᵐ

FIRST TV PICTURES

INFRARED
TEMPERATURE
SWEEP

RADIO BEAMS TO AND
FROM EARTH THROUGH
ATMOSPHERE

CANOPUS

Figure 4.2   Path of Mariner 10 shows how the spacecraft approached Venus from the night side of the planet, was then hidden from the Earth during occultation when the path of the radio link penetrated the atmosphere, and continued away from Venus taking pictures of the sunlit hemisphere of the planet.

9:21 A.M. PDT   Television cameras start photographing the dark limb of the planet. The pictures are blank as expected. Still Mariner 10 flies on undisturbed.

9:49 A.M. PDT   The first TV picture showing a lighted cusp of Venus.

A few minutes later this first picture of Venus (Fig. 4.3) is displayed on television monitors at the Jet Propulsion Laboratory. Taken at a distance of 5,000 miles (8,000 km) from Venus, the picture reveals a thin bright sliver of the north pole of Venus, the horn of an extremely fine crescent. At once it is clear that there are none of the irregularities that several Earth-based observers had claimed to see on the horns of the crescent Venus and postulated as being due to clouds. The cusp appears feature-less and geometrically perfect.

62

Figure 4.3   Mariner 10's television cameras took this first wide-angle picture of Venus at 9:49 A.M. PDT on February 5, 1974. The photo shows the lighted cusp of Venus near the north pole. Mariner 10 was approaching Venus from the night side at more than 20,000 miles per hour (32,000 km/hr). The picture was taken from about 5,000 miles (8,000 km).

10:01 A.M. PDT   Mariner makes its closest approach to Venus, hurtling 3,600 miles (5,800 km) above the cloud tops.

With 75 percent of the illuminated disc of Venus now in view of the spacecraft and Mariner 10 still firmly locked on Canopus, everyone begins to feel more confident. The critical point has been passed. Mariner 10's star sensor has not been distracted by the brightness of the planet. The TV cameras continue to take pictures. As these are displayed on the monitor screens at JPL, they show somewhat disappointing but not entirely unexpected vistas of foglike clouds devoid of any visible detail. There is no structure whatever to the top of the visible cloud deck. Venus's clouds are certainly not like Earth's (Figs. 4.4 and 4.5).

10:07 A.M. PDT   Mariner 10 enters into Earth occultation as it passes on the far side of Venus and the bulk of the planet comes between the spacecraft and Earth.

Figure 4.4 *THE CLOUDS OF EARTH.* Seventeen hours after launch, Mariner 10 acquired this view of the Earth from a distance of 124,000 miles (200,000 km). North is at the top, and the illumination is from the left. The frame is 1,800 miles (3,000 km) wide and shows a scattered cloud formation over the Pacific Ocean and Central America. Towering cumulonimbus clouds can be recognized in the center right-hand portion where they are highlighted by the oblique sunlight striking their protruding surfaces. Compare this picture with Figure 4.5.

Radio signals from the spacecraft begin to fade at the Goldstone receiving antenna as the occultation experiment commences. But data and pictures are still being gathered by the spacecraft. These are stored in its tape recorder for later transmission to Earth after the occultation. The occultation experiment performed by Mariner 10 is much more sophisticated than that of Mariner 5. The higher frequency experimental X-band provides the second frequency this time, affording the possibility of detecting finer detail in the atmospheric structure. Most important, commands to the high-gain dish antenna on Mariner 10 cause it to turn slowly and point *away* from the Earth—just enough to compensate for the increasing refraction caused by the dense Venusian atmosphere and to probe longer into the atmosphere. A few minutes later the spacecraft begins to emerge from occultation. As a result of this precisely programmed operation by an automated spacecraft, carried out during the critical time when the spacecraft's vital communications with Earth are being interrupted, new information about the internal structure of Venus's atmosphere is obtained.

Figure 4.5 *THE CLOUDS OF VENUS*. Seventeen hours after its closest approach to Venus, Mariner 10 acquired this view from a distance of 317,000 miles (510,000 km). The frame covers a region about 3,100 miles (4,500 km) wide and was acquired in blue light. North is at the top. Illumination is from the left. The picture has been enhanced to bring out any small-scale detail resulting from discrete cloud structure at the top of Venus's atmosphere analogous to that seen in the Earth picture (Figure 4.4). It is obvious that there is no such structure visible in this picture down to a resolution of about 4.3 miles (6.5 km), nor was any found in other frames which could have detected vertical structure of only a few tens of meters relief. The patterns of dark dots seen in this frame and in Figure 4.4 are calibration marks on the faceplate of the vidicon camera.

A most important contribution to planetary atmospheric sciences is yet to come. The cameras have been switched to ultraviolet filters during the encounter. As these high-resolution ultraviolet pictures are viewed on Earth, the atmosphere of Venus is seen to be a panorama of detailed features never before recorded (Figs. 4.6 through 4.11) because Earth-based ultraviolet pictures show only the grossest features. For reasons still not well understood, cloudlike markings at the top of the Venusian atmosphere are manifested in ultraviolet light but not in the nearby blue portion or any other part of the visible spectrum. Mariner 10's great discovery is that these markings act to some extent as tracers of atmospheric motion. The carefully planned close-up mosaics acquired by Mariner 10 prove to be incredibly detailed "snapshots" of the top of the atmosphere, displaying a highly organized system of motion quite different from Earth and Mars. Taken in sequence, these snapshots provide a time-lapse film of Venus's atmosphere in motion (Figs. 4.12 through 4.16). Indeed, such a motion picture has been prepared from

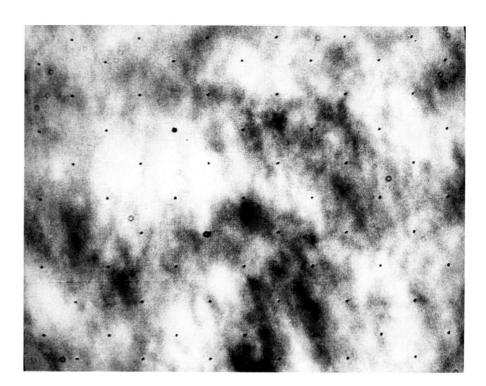

Figure 4.6 *PANORAMA IN "BLACK" LIGHT, I.* In contrast to the disappointing featureless views seen in visible light, there is a surprising pattern of fine detail in the pictures of Venus obtained when ultraviolet filters were used at an effective wavelength of about 3,550 angstroms, which permitted only light in the invisible wavelengths beyond the blue to be recorded by the vidicon tube. This frame covers a region about 930 miles (1,500 km) square, with a resolution of about 6.8 miles (11 km). It is centered at approximately 11°N latitude and has been enhanced by computer processing. The equal-dimensional light and dark features observed here may be associated with large-scale convection cells present within the upper troposphere and lower stratosphere of Venus. The reasons that atmosphereic motions and/or temperature differences are displayed so strongly in ultraviolet light and are completely lacking in the visible range have not been satisfactorily determined.

Figure 4.7 *PANORAMA IN "BLACK" LIGHT, II.* A generally similar appearance to that shown in Figure 4.6 was observed at the slightly higher resolution of 5 miles (8 km) almost exactly on the equator in the view shown here. The processing reveals the details of the ultraviolet scattered back by the atmosphere of the planet and also shows several small and normally inconspicuous aberrations of the camera itself. The doughnut-like circles seen in several places on the picture are due to dust particles near the focal plane of the vidicon tube.

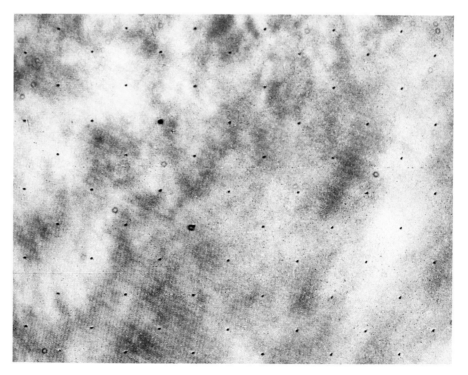

Figure 4.8 *PANORAMA IN "BLACK" LIGHT, III.* The character of the ultraviolet markings begins to change as the photography is extended farther away from the equator. This view, located at 23°S latitude, covers an area about 1,250 miles (2,000 km) square at a resolution of about 4.3 miles (7 km). The pattern of markings tends to be organized in streaks or bands that originate in the equatorial regions and generally seem to spiral to the higher latitudes. In addition, faint banding or wave structure opposite to the main direction of light and dark markings can be recognized.

Figure 4.9 *PANORAMA IN "BLACK" LIGHT, IV.* As the atmosphere is examined progressively farther from the equator, the streaky and banded structures become more apparent. In this frame, centered at 36°S latitude and covering an area 1,120 miles (1,800 km) square, the resolution is about 3.7 miles (6 km). Like all the other ultraviolet frames in this series, it has been computer enhanced to exaggerate artificially the intrinsic scene contrast.

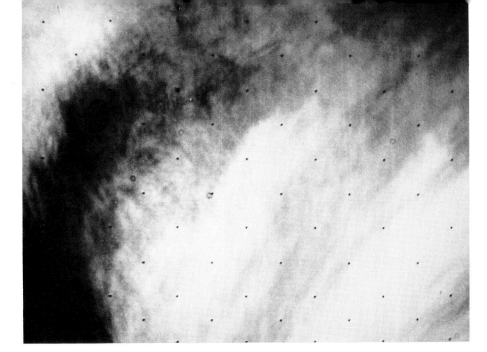

Figure 4.10 *PANORAMA IN "BLACK" LIGHT, V.* By one day after Mariner 10's approach to Venus, individual TV frames began to record larger-scale atmospheric cloud patterns. In this ultraviolet-filter frame, centered at 11°S latitude, the region covered is 5,000 miles (8,000 km) across. The subsolar point, i.e., the point on the planet at which the Sun is vertical, is located in the left-central portion of the frame, and the equator of Venus slants from lower left to upper right nearly through the middle of the frame. It became evident from these pictures that there is a tendency for an equal-dimensional, perhaps convective, pattern to be displayed in the equatorial regions, especially near the subsolar point. In contrast, streaky, organized, spiral structures are seen at the higher latitudes (Fig. 4.11). At the upper boundary of the dark patch in the upper left-hand portion of this frame can be seen a round, light-toned structure with a dark center. It is about 174 miles (280 km) across. In repetitive photography such features were observed to change character or disappear entirely within a matter of several hours, suggesting rapid rates of vertical and horizontal motion in the upper troposphere and lower atmosphere where these markings originate.

Figure 4.11 *PANORAMA IN "BLACK" LIGHT, VI.* The southern polar region and limb area are seen in this ultraviolet photograph of Venus taken 21 hours after encounter. The solid white ring is about 435 miles (700 km) wide, and it has been speculated that this is the site of a vortex structure fed by meridional flow from the equatorial regions at high elevation balanced by descending air on the south polar regions that returns to the equator at greater depths within the atmosphere. Illumination is from the left, and the terminator of Venus is along the right-hand side.

the Mariner 10 pictures. The suspected four-day rotation is found to be the average motion in the equatorial regions. Much higher speeds are discovered in the higher latitudes, with even a suggestion of a gigantic atmospheric whirlpool surrounding each pole.

Venus's atmosphere is truly unearthly, not only in composition but in dynamics. The atmosphere is so opaque to sunlight that most of the absorbed solar energy and consequent heating take place near the top of the atmosphere rather than at the bottom as on Earth and Mars. Venus, our closest planetary relative in size, mass, and composition, manifests a bizarre and uninhabitable surface environment. Yet we Earthlings should examine carefully this aberrant relative. Had Earth been located 70 percent closer to the Sun, at the orbit of Venus, it too would probably be uninhabitable and uninhabited, so delicate is the balance of absorption and reradiation of the Sun's energy.

To avoid accidentally destroying this delicate balance in Earth's atmosphere by mankind's activities, it is mandatory that we understand fully the ecosphere in which we have evolved. This understanding is difficult, if not impossible, without comparisons to other planetary atmospheres. The new information provided about the atmosphere of Venus by Mariner 10 is important to planning further investigation of that atmosphere by space probes.

Figures 4.12 and 4.13 are on following pages.

Figure 4.12 *THE GLOBAL VIEW.* This is a reproduction of a photomosaic of ultraviolet pictures of Venus acquired about 6½ hours after encounter and has in the original version a resolution of about 5.2 miles (7 km). Venus has a diameter of 7,521 miles (12,104 km). The pictures have been reconstructed to display approximately the range of intensities actually exhibited on Venus. For that reason the area near the terminator is quite dark and causes a less than circular overall appearance. The bright polar ring in the south pole is conspicuous in this frame, as is the dark convective zone near the subsolar point. Dark streaks emanate from the equatorial regions and spiral up toward the polar region. The equator runs approximately through the center of the dark, equatorial band. The features displayed in the photomosaic appear to be closely symmetrical about the equatorial plane.

Figure 4.13 *THE GLOBAL VIEW, ENHANCED.* This is the same set of picture data used to produce the mosaic in Figure 4.12, but processed by computer to enhance small-scale detail. Such processing necessarily suppresses large-scale variations in brightness in order to produce an artificial image in which visible detail corresponding to as little as 1 or 2 percent intensity variation can be displayed. The spiral structure is much more apparent than in Figure 4.12, and structure can be recognized within the south polar bright ring itself. In addition, the detailed structure in the equatorial regions shows up more clearly. Such a view represents a double extension of human vision, first, by displaying an image in the invisible portion of the spectrum beyond the blue, and second, by conspicuously displaying gray tones that are intrinsically too subtle to be resolved even by special film, much less by the human eye.

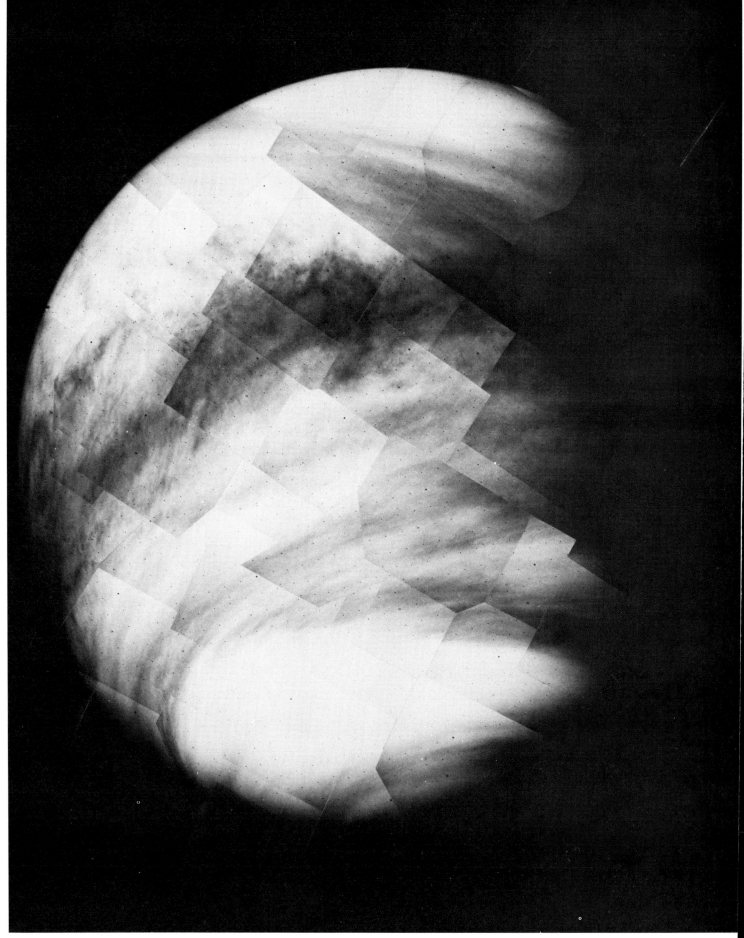

4.12

4.13

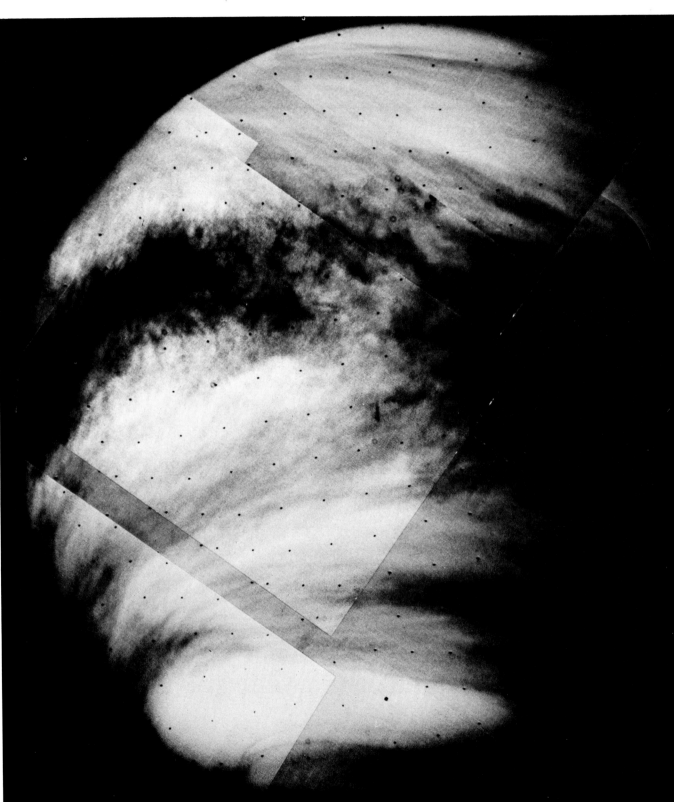

Figure 4.14 *THE ROTATION OF VENUS, I.* The photomosaics of Figures 4.14a and 4.14b were taken four hours apart, approximately one day after the encounter. Careful comparison of the structure in the central region of the mosaics where the dark equatorial ring exists should satisfy the reader that tangible rotation from right to left was recorded— that is, in a retrograde direction, opposite to the rotation of the Earth and Mars. The rotation of such cloud features has been determined to correspond to about one rotation every four Earth-days, which corresponds to a speed of about 330 feet per second (100 meters/sec). The solid surface of the planet also rotates in a retrograde sense, but much more slowly—once every 243.1 days. Thus the top of the atmosphere of Venus rotates about 60 times faster than the surface.

a

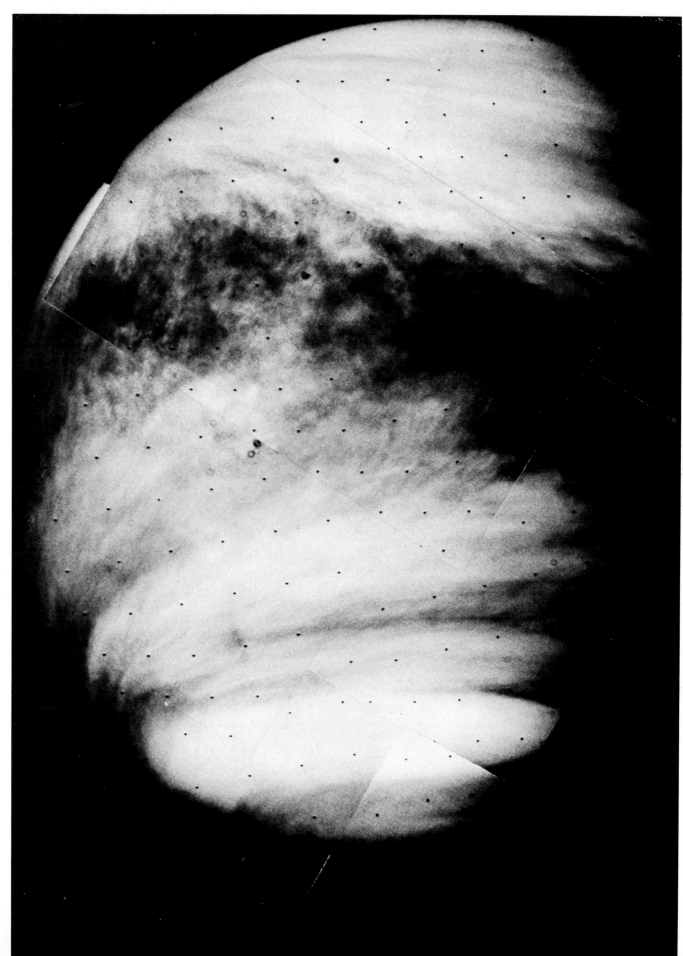

b

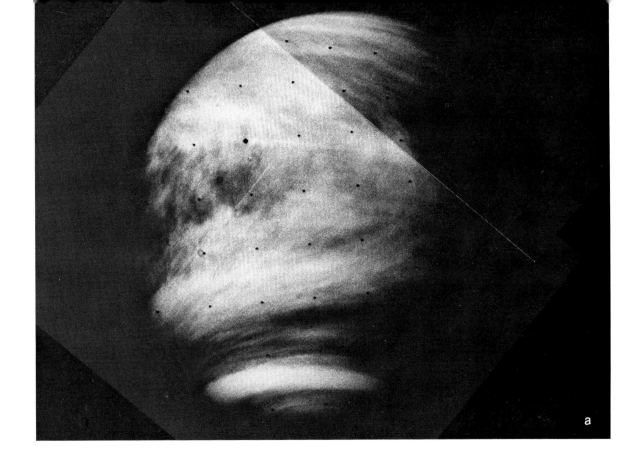

a

Figure 4.15 *THE ROTATION OF VENUS, II.* These four mosaics, taken at 110-minute intervals (displayed in the order *a, b, c, d*) were acquired approximately 57 hours after encounter. Resolution is about 30 miles (50 km). Structure in the dark equatorial band can be seen to rotate from right to left, consistent with the four-day rotation period discussed with Figure 4.14. The rapid rotation of the top of the atmosphere as observed in these ultraviolet pictures, compared to the very slow rotation of the planet's surface, has been attributed to the fact that Venus's atmosphere, unlike that of Earth or Mars, is heated

b

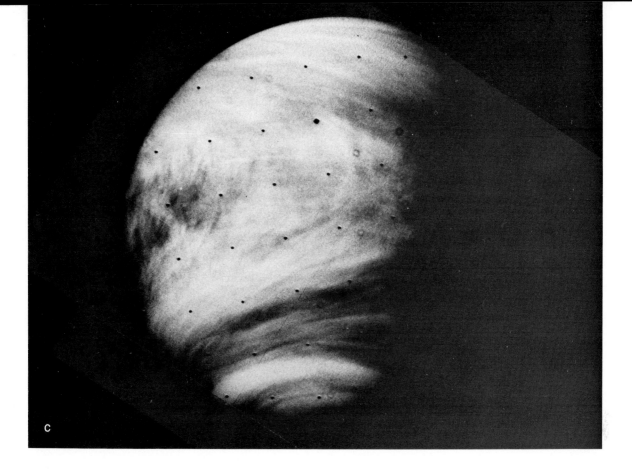

c

from the top because of its great opacity. An unusual circulation pattern develops, involving very rapid rotation of the upper part compared to the lower. In addition, a sub-solar disturbance area develops, in the region where the Sun is shining vertically down-ward over the planet, and stays locked to the direction of the Sun. The average zonal (east-west) flow must pass around and through this disturbed region, perhaps creating some north-south atmospheric circulation that in turn produces streaks evident in the mid-latitude and polar regions.

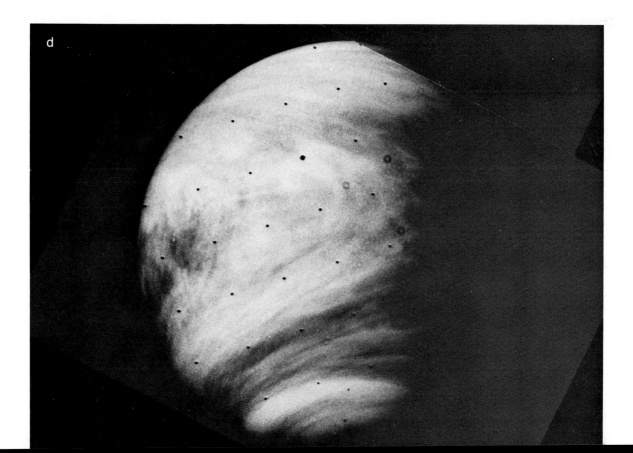

d

 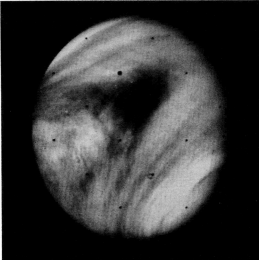 

+ 3 days **+ 4 days** **+ 5 days**

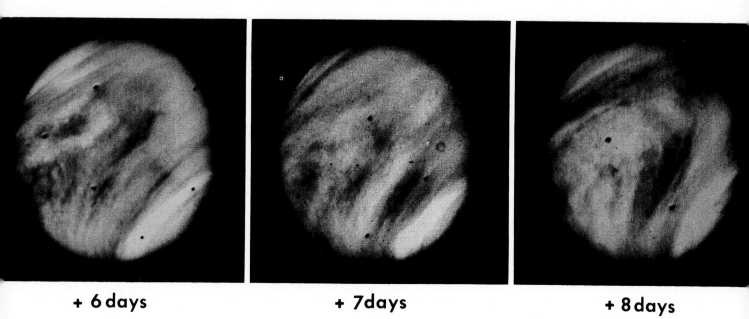

+ 6 days   + 7 days   + 8 days

Figure 4.16 *THE ROTATION OF VENUS, III*. In this set of frames the appearance of the planet over a six-day interval is monitored. The general configuration of the markings remains surprisingly similar even though individual features, especially in the equatorial regions, move at the characteristic four-day rotation rate.

5

Mercury Revealed

Mercury is of particular interest as the end member of the sequence of terrestrial planets; some estimate of its internal activity may be obtainable from Mariner-type surface photography. Although there is little prospect that Mercury has its own magnetic field, a measurement thereof should be attempted on a flyby.
—Space Science Board, National Academy of Sciences, 1968

February 13, 1974   Venus picture-taking is concluded after 4,165 planetary images have been obtained.

Mariner 10's encounter with Venus has been an unqualified success, far exceeding expectations. But Mercury, the main objective, still lies ahead. Serious spacecraft problems remain unsolved: the antenna power is still down; the mysterious oscillation problem with the attitude control system remains unsolved; the supply of the precious attitude control gas is low.

February 13, 1974   NIXON REFUSES TO TURN OVER MORE WATERGATE TAPES TO SPECIAL PROSECUTOR JAWORSKI

February 13, 1974   NOBEL PRIZE–WINNING AUTHOR AND POLITICAL DISSIDENT ALEXANDER SOLZHENITSYN IS EXILED FROM HIS NATIVE RUSSIA

The spacecraft is now readied for the third trajectory correction maneuver (TCM), essential for a close pass of Mercury. Minutes before critical tests are to start, the radio command transmitter at Goldstone loses contact with the spacecraft. The tests and the TCM have to be delayed to the following day.

Mariner 10's gyros are tested on February 14, and the system behaves properly during the first two tests. During the third test, however, uncontrolled oscillations of the roll gyro system are encountered. They also occur during two more tests using commanded turns. About 7 percent of the remaining nitrogen attitude-control gas is thus used up. The oscillations resulted not from a fault in the gyro system as such, but from structural interaction between the spacecraft and the gyro system, a resonance effect like pushing someone on a swing or like a violin note breaking a crystal glass.

The planned TCM is canceled; it is just too risky with the roll oscillation problem. Instead, a novel "sun-line" maneuver is worked out which can be delayed for about a month. It would make use of the natural configuration of Mariner's orbit around the Sun and the orientation of the spacecraft in space. The maneuver was fortuitously made possible because the small error of 10.6 miles (17 km) in the Venus flyby point was in the right direction. By mid-March, Mariner 10 will be in a location and orientation such that, without pitching or rolling the spacecraft, its rocket engine will point in just the right direction in space to provide a thrust that will change the trajectory to pass through the desired location off the dark side of Mercury on March 29. The roll oscillation problem can thus be circumvented by not using commanded roll turns for the required TCM.

Again, all seems well. The ingenuity of the project personnel has overcome the deficiencies of the spacecraft's roll turn capabilities and has developed extemporaneously a technique to get the spacecraft to its rendezvous with Mercury. But had the sun-line maneuver not been available, a very hard choice would have had to be made: risk all for a good aim point and chance of return to Mercury for a second encounter, or accept a less desirable encounter with no chance of a return.

February 18, 1974   Shortly after midnight, Mariner's star tracker loses its lock on Canopus. The gyros come on and the spacecraft begins an automatic search for Canopus. The star tracker "sees" a bright nearby particle—possibly a speck of dust traveling alongside the spacecraft. Brightly illuminated by the Sun, this particle looks like a star to the automatic tracker. The spacecraft is tricked into tracking the particle instead of Canopus. Misoriented, the spacecraft no longer points the high-gain antenna toward Earth! Communications are interrupted because the smaller ground antenna then in use cannot communicate through the low-gain antenna of the spacecraft.

When effective communication is restored through use of a large ground-based antenna, a command is sent to the spacecraft to start a normal roll search, and Canopus is reacquired within 1.3 minutes. This bright particle distraction not only placed the spacecraft in a disoriented situation but also caused a loss of more maneuvering gas.

February 19, 1974   FEDERAL ENERGY OFFICE ORDERS EMERGENCY GASOLINE
                     ALLOCATION TO 20 STATES

February 25, 1974   HERBERT KALMBACH, NIXON'S PERSONAL ATTORNEY, PLEADS
                     GUILTY TO CRIMINAL CHARGES OF ILLEGAL CAMPAIGN
                     ACTIVITIES

On February 27, the Mariner star tracker receives another of the frequent disturbances from bright particles. However, this is only a momentary distraction, and the tracker reacquires Canopus before Mariner 10's gyros turn on. The bright particles have been disturbing Mariner since launch. Initially, they occurred only once or twice a week; now the distractions come more frequently—as many as 10 per week, possibly because the small dust particles traveling along with Mariner on its voyage to Mercury reflect more light as the intensity of sunlight increases.

February 28, 1974   Mariner 10's solar panels are tilted another 10 degrees away from the increasing solar radiation to keep them at the right temperature. Total tilt is now 68 degrees from their original position when extended after launch. They are still slightly too warm, so the following day the solar panels are tilted an additional 10 degrees.

March 4, 1974   Good news for a change! The radio signal being transmitted by Mariner 10's high-gain antenna suddenly and inexplicably increases to normal. Project management had hoped this might happen when a change in solar illumination of the antenna occured at this time. Full-strength transmission is essential at Mercury to permit the hoped-for 117-kilobit data rate necessary to return the planned high-resolution pictures of the planet. If the antenna performs as it is designed to do, *five times* more high-resolution pictures will be received than would be returned to Earth if the antenna problem had continued.

As one problem with the neurotic spacecraft clears, another seems to develop. On March 6, a group of bright particles again disturb the star tracker. The spacecraft goes into an uncommanded roll, and the gyros automatically turn on to correct it. They remain on for about 40 minutes. Fortunately, the earlier oscillation problem does not occur, so that only a small quantity of maneuvering gas is lost. But every millipound is now beginning to count. Something must be done before a real catastrophe takes place that could make Mariner 10 into a derelict spacecraft before it gets to Mercury.

As a result of long, anguished conferences, project management decides to reprogram the spacecraft on March 8 so that it will not turn on its gyros automatically. In the future, should a bright particle distract the Canopus sensor and the spacecraft begin to roll, Mission Control, using the engineering measurements radioed back constantly, will decide

81

when the gyros should be turned on. Thereby the use of maneuvering gas will be closely controlled. This procedure involves yet another calculated risk, because without the automatic response capability some damage to the spacecraft might take place before ground controllers could decide how to intervene effectively and transmit corrective commands to it. The risk, however, is deemed to be very small.

Most people go home from the Jet Propulsion Laboratory to enjoy a southern California weekend, but events are now so critical with Mariner 10 that a team of spacecraft experts work straight through for several weekends—developing a new "energy-conserving" technique of navigating Mariner 10 by solar sailing.

o   o   o

"A sailboat trip around the Moon propelled by solar radiation could provide a challenge for next generation's teenagers." Maurice D. Arnold, Regional Director of the Bureau of Outdoor Recreation, was addressing a meeting of the American Astronautical Society in Denver, in July 1969, when he jocularly suggested such an adventure. He was referring to the fact that the intense flow of radiation from the Sun had been found to exert a real force on objects such as satellites in space.

Actually, the concept of solar sailing dates back to 1946, when an Italian, Luigi Gussalli, wrote a slim volume called *I Viaggi interplanetari per mezzo delle radiazioni solari.* In this book he suggested that dust particles could be accelerated by solar light pressure to drive a spacecraft as a gust of winds on sails. He outlined a complicated scheme to augment the dust in space and effectively increase the strength of his space wind.

o   o   o

Now the Mariner project will use solar sailing in earnest. The tilt of Mariner's solar panels is arranged so that the torque produced on them by solar radiation will produce a roll counter to the natural drift in the roll of the spacecraft. "Solar sailing" as applied to Mariner 10 is really a misnomer, since the effect is being used to assist in keeping the spacecraft oriented correctly rather than to derive a propulsive "sailing" effect as envisioned by Gussalli and Arnold. But now Mariner can be maintained in a stable condition through slight changes of the tilt of the solar panels, supplementing cycling of the maneuvering gas jets that maintained the roll position in the normal cruise mode of the spacecraft with the high-gain antenna pointed toward Earth. By March 12, the spacecraft, despite particle distractions, is in this normal cruise condition with its roll attitude stabilized. Consumption of attitude stabilization gas is minimal, and the danger of uncontrolled disorientation of the spacecraft is greatly lessened. Creative problem-solving decreases the danger of a failure before reaching Mercury and later will prove to be an essential step in making possible two additional visits to Mercury by Mariner 10.

March 16, 1974
4:54 A.M. PDT
The planned sun-line TCM is attempted. Without any need for change in Mariner 10's orientation, the rocket engine on the spacecraft is commanded to burn for 51.1 seconds. This thrust produces a velocity change to the spacecraft of about 59 feet per second (18 meters/sec) directly away from the Sun—the first sun-line maneuver ever carried out in the Mariner program. Innovation again pays off as the path of Mariner 10 shifts just enough for the desired near-miss off the dark side of Mercury (Fig. 5.1).

March 18, 1974
ARAB OIL BOYCOTT ENDED, BUT PRICES REMAIN UNCHANGED

March 19, 1974
CONSERVATIVE SENATOR JAMES L. BUCKLEY CALLS FOR NIXON'S RESIGNATION

March 24, 1974
Mariner 10's first picture of Mercury is acquired from a distance of 3,340,000 miles (5,380,000 km). Even with computer enhancement, pictures taken during the next few days show very little detail—just the indistinct shadings that have been seen so long from Earth (Fig. 5.2).

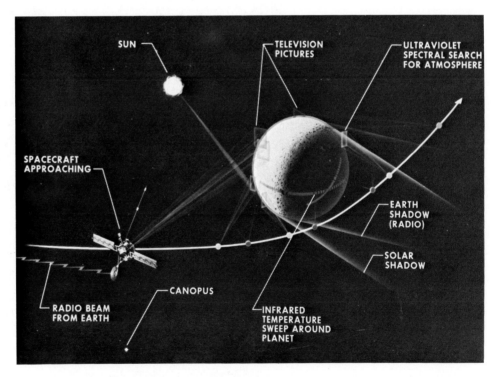

Figure 5.1   The path of Mariner 10 past Mercury at the first encounter brought the spacecraft to within 470 miles (756 km) over the night side of the planet. At this time the spacecraft was also hidden from Earth by the disc of Mercury and was thus out of communication with Earth.

83

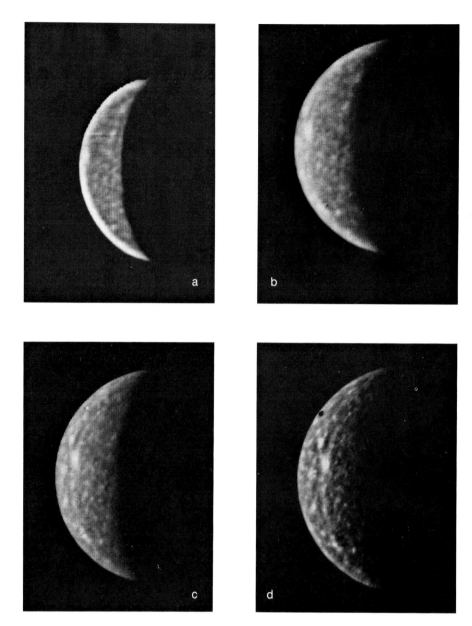

Figure 5.2 *ZOOMING IN ON CRATER KUIPER, I.* These computer-enhanced frames illustrate the appearance of Mercury as it was first approached by Mariner 10; the limb or edge of the planet is to the left, the terminator or boundary between night and day is to the right of each image, north is at the top, and illumination is from the left. The bright surface feature in the center left-hand side is the crater Kuiper with its surrounding blanket of ejecta—one of the brightest regions on the entire visible hemisphere of Mercury observed by Mariner 10. The first frame at left was taken 6 days from closest approach at a range of 3.34 million miles (5.38 million km) and has a resolution comparable to the best Earth-based telescopic views (see Fig. 1.1b). The second frame was taken five days before closest approach at a range of 2.8 million miles (4.5 million km); the third frame, four days before at a range of 2.27 million miles (3.65 million km); and the fourth frame, three days before closest approach at a range of 1.8 million miles (2.75 million km). The prominent black dot in the fourth frame is a calibration marking on the vidicon faceplate of the television camera. The resolution of the fourth frame is 47 miles (75 km). For scale, the diameter of Mercury is 3,030 miles (4,878 km).

If imagery of Mercury at 2-kilometer resolution is indeed obtainable . . . it becomes a most significant experiment from the viewpoint of planetary surfaces. . . . The best way to underscore the scientific value of [even] a limited imagery mission for Mercury is to recall the major changes in our thinking about Mars produced by the 4-kilometer resolution Mariner 4 pictures.—National Academy of Sciences, Space Science Board, 1968

The Jet Propulsion Laboratory becomes a mecca for many planetary scientists as the day of encounter nears. Science correspondents from the international and national press converge on the Von Karman Auditorium at the Laboratory for their first views of the innermost world. There is an air of expectancy everywhere.

A picture taken at 2,190,000 miles (3,513,000 km) is enhanced by the computers at JPL and shows a mottled surface to Mercury. The planet appears as a 6-day-old crescent moon. This is the first picture that suggests a heavily cratered surface, like the Moon's. Also, a large bright feature is seen near the limb, but it is still too indistinct to be identified as a depression or a mountain (Fig. 5.2).

At a distance of 1,160,000 miles (1,870,000 km), on March 27, Mariner is producing pictures that now clearly show craters dotting Mercury, many of them huge (Fig. 5.3). The bright spot looks more and more like a bright-rayed crater, and there are also light streaks along great circles, reminiscent of some of the lunar rays.

About 144 pictures of Mercury have now been returned to Earth. The best ones so far are showing features as small as 25 miles (40 km) across. The imaging mission to Mercury appears to be approaching phenomenal success. Each day, then each hour, brings new vistas of a heretofore unseen planetary surface (see Figs. 5.4 to 5.17).

As Mariner 10 approaches Mercury, other scientific observations can be made that may provide information about the interior of the planet. Three experiments on particles and fields are directed to finding out how Mercury interacts with the interplanetary medium, particularly the solar wind. These experiments are plasma science, charged particles, and magnetic fields.

The plasma science experiment measures natural electrical currents in space and has already made the first observations of the solar wind inside the orbit of Venus. The interaction of the solar wind with Earth, Moon, and Venus and Mars has been found by spacecraft experiments to be quite different. It is generally believed that interaction with Mercury will be very similar to that with the Moon, where the solar wind impinges directly on the planetary surface and the bulk of the planet causes a cavity in the wind trailing behind the planetary body. This is

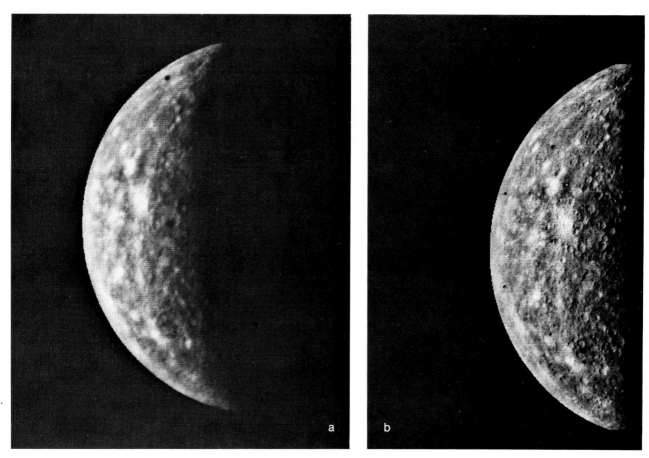

Figure 5.3 *ZOOMING IN ON CRATER KUIPER, II.* These photographs were taken at 48 hours (left) and 24 hours (right) before closest approach and illustrate the effect of increasing surface resolution from 30 to 15 miles (50 to 25 km). The spacecraft was 1,160,000 and 590,000 miles (1,870,000 and 950,000 km) from the planet, respectively. The set of six frames of this figure and Figure 5.2 illustrate the effect of the rotation of Mercury during the six days of approach to the planet. The bright spot of crater Kuiper conspicuously moves from left to right as the planetary rotation of approximately two Mercurian "hours" takes place. If Mariner 10 had arrived at Mercury four days later, crater Kuiper would never have been seen at high resolution. Also, note in the first and second photos of Figure 5.2 two other small bright areas near the terminator, to the north of crater Kuiper. They are rotated into the terminator in the closer photographs. Presumably they are bright craters as well. At high resolution the curved bright streak passing north-south through crater Kuiper can be seen to be a bright ray ejected from the crater itself. More black calibration dots are visible on these two frames.

very different from Earth and Venus and Mars, where the solar wind is held away from the surface of the planet—from Earth because of its magnetic field, and from Venus and Mars because of their ionospheres. The path of Mariner 10 is expected to carry it through the anticipated plasma cavity behind Mercury.

Unfortunately not all of the plasma experiment can be performed be-

Figure 5.4 *ZOOMING IN ON CRATER KUIPER, III.* Fourteen and a half hours before closest approach, Mariner 10 acquired this view centered on the crater Kuiper region from a distance of 342,000 miles (550,000 km). Resolution is about 12.5 miles (20 km). The largest craters in the photo are about 75 miles (120 km) across. Crater Kuiper itself can now be resolved as a fresh, bright crater lying on top of older craters and covering them with its ejecta materials.

cause a protecting door has refused to open since the launch, but the secondary plasma experiment, a rearward-looking electrostatic analyzer, functions well. This analyzer has been measuring the number (flow) and energy (temperature) of solar wind electrons as Mariner 10 approaches Mercury. All appears to be going according to theory as the spacecraft bears down on the planet. The interplanetary rate of electrons per second (flux) continues. The solar wind is undisturbed by Mercury. Now the spacecraft is about 20 minutes before closest approach. Experimenters watch for the telemetered data to drop suddenly toward zero flux as the spacecraft passes into the expected plasma cavity behind Mercury.

Instead, at 19 minutes before closest approach, the plasma flux suddenly jumps and peaks in a way that suggests Mariner 10 has crossed a bow shock wave. The scientists are astounded. How could a slowly rotating, seemingly Moon-like planet such as Mercury, probably devoid of any atmosphere, give rise to a bow shock in the solar wind the way Earth and Jupiter with their magnetic fields do?

87

5.5

Figure 5.5 *ZOOMING IN ON CRATER KUIPER IV.* Five hours before closest approach, Mariner 10 recorded this picture from a distance of 112,000 miles (180,000 km). The resolution is 5.28 miles (8.5 km). Crater Kuiper, which is about 25 miles (40 km) across, is clearly resolved, nested in the edge and rim of an older 50-mile (80-km) diameter crater. Other smaller, bright-rayed craters are becoming more visible as the resolution increases. The serrated appearance of the limb in this and some of the preceding frames is an artifact of the reconstruction of the pictures.

Figure 5.6 *ZOOMING IN ON CRATER KUIPER, V.* As Mariner 10 approached Mercury at nearly 7 miles per second (11 km/sec), the TV cameras took this picture of crater Kuiper from a distance of 53,000 miles (85,000 km). Now the ejecta blanket of the older larger crater can also be seen along with chains of secondary craters extending out radially for about 63 miles (100 km). Secondary craters are caused by large blocks of rock thrown out by the impact that produces a primary crater. How far from the primary the secondaries are spread depends upon the energy of impact and the surface gravity of the planet. On Mercury, for example, secondary craters are closer to their primaries than are those on the Moon. The radial grooves are likewise caused by superimposed secondary impacts. Extensive areas of intercrater plains can also be recognized, delineating the large craters from one another.

Figure 5.7 *ZOOMING IN ON CRATER KUIPER, VI.* This enlargement of the previous frame illustrates the same basic 1.86-mile (3-km) surface resolution.

88

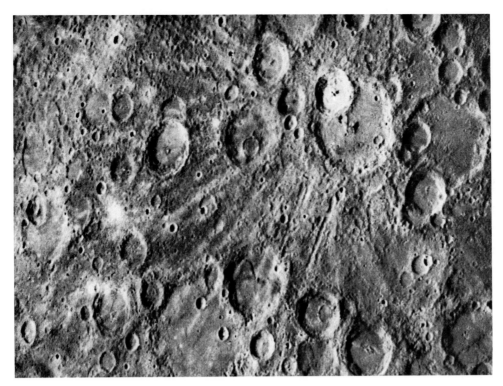

5.6

5.7

Figure 5.8 *THE INCOMING HEMISPHERE OF MERCURY.* This photomosaic is assembled from individual high-resolution Mariner 10 frames acquired shortly before closest approach to Mercury. The terminator is at the right, the limb at the left, and the illumination by the Sun is from the left. North is at the top, and the equator arcs from left to right about two-thirds of the way up the disc. The terminator is located at about 10° W longitude. Crater Kuiper can be recognized just above the center of the illuminated half-moon. The landscape is dominated by large craters and basins with extensive intercrater plains areas as well. The white rectangle outlines an area of a very peculiar terrain, enlarged in Figure 5.9.

Figure 5.9 *PECULIAR TERRAIN, I.* In this enlargement of Figure 5.8 a large, smooth-floored crater of 106 miles (170 km) diameter is centered. It is also shown in Figure 5.10. North is at the top. Sun illumination is from the left, as in the preceding pictures. A long, northwest-trending valley from the center crater extends more than 60 miles (100 km) and has been named Arecibo Valles. (The world's largest planetary radar dish is located at Arecibo, Puerto Rico. It was used to measure for the first time the correct rotation period of Mercury.) The valley is about 4.3 miles (7 km) wide. This peculiar terrain covers a semi-elliptical area of at least 193,000 square miles (500,000 sq. km) centered at 20°S latitude and 20°W longitude. Approximately antipodal to this region, on the other side of the planet, is a very large basin of over 800 miles (1300 km) diameter, comparable to the Imbrium Basin of the Moon.

90

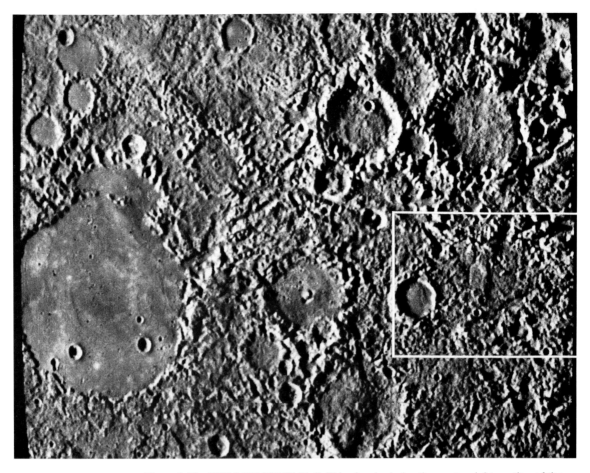

Figure 5.10 *PECULIAR TERRAIN, II.* This view includes the center right portion of the photomosaic shown in Figure 5.9, again showing the large, smooth-floored crater. The photograph covers an area extending approximately 310 miles (500 km) in horizontal dimension, and the resolution is better than 1.2 miles (2 km). Note the very smooth, flat-floored crater bottoms as compared to the badly degraded rims and intercrater areas. The plains that fill the crater floors are obviously younger than surrounding terrain and did not suffer the disruptive process that destroyed previous crater forms. The small, smooth-floored crater in the center right portion (included in the area outlined in white) is shown in Figure 5.11.

Meanwhile, the charged particle experiment is also producing unexpected data. This experiment uses charged particle telescopes that detect cosmic rays—atomic nuclei speeding through space with a wide range of energies and atomic number—and high-energy electrons. During Mariner's cruise between the planets, the instrument has measured both solar and galactic cosmic rays.

The interplanetary levels of energetic charged particles remain normal until Mariner 10 is about 20 minutes before its closest approach. Again the counting rate jumps violently, signifying that something is abruptly

Figure 5.11 *PECULIAR TERRAIN, III.* Taken from a range of 11,200 miles (18,000 km), this is a high-resolution view of the outlined area seen in Figure 5.10. The terrain consists of numerous dissected hills from a few hundred yards to over a mile high (a few hundred meters to 2 km), interspersed with smooth material. The crater in the left-center, its floor peppered with small craters, is 19 miles (31 km) in diameter. It has been speculated that this hilly and lineated terrain was produced as the result of enormous crustal disturbances associated with the focusing of seismic energy from a cataclysmic collision on the opposite side of the planet which produced a very large impact basin there. The smooth plains material filling these large craters is presumably of younger and probably volcanic origin.

changing the solar wind. The increase in numbers of charged particles is interpreted as being caused by Mariner crossing a bow shock wave. But beyond this crossing there are several unusual peaks of intensity of the particles that appear to be evidence of a pulsating acceleration mechanism which rapidly boosts the energies of solar wind protons and electrons from a few electron volts to hundreds of thousands of electron volts. The peaks occur too rapidly to represent belts of trapped particles as are found surrounding the Earth. They are reminiscent of phenomena that have been observed in the geomagnetic tail trailing behind the

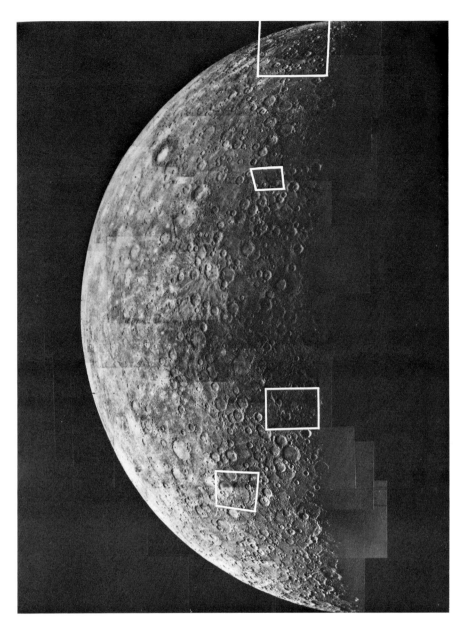

Figure 5.12 *THE SCARPS OF MERCURY.* Quite distinct from Mars and Moon, the cratered terrains of Mercury display prominent scarps, or steep slopes, up to 1.8 miles (3 km) high and hundreds of kilometers in length. They generally exhibit a curved or slightly lobate (scalloped) appearance and face eastward—suggesting that the planet shrank and caused compressional faults on the surface. Four of the most conspicuous are shown in the succeeding frames. They are Victoria, Santa Maria, Discovery, and Vostok. Scarps on Mercury are named after ships of discovery.

Figure 5.13 *VICTORIA SCARP.* In this view of Mercury's northern limb taken from 48,300 miles (77,800 km), the prominent east-facing scarp extends from the limb southward for hundreds of kilometers. The horizontal "tear" in the picture was caused by a momentary loss of data. The dimension shown is about 360 miles (580 km) across. Illumination is from the left. The scarp is named after Magellan's ship, the first to sail around the world.

Earth, where oppositely directed field lines meet. High-energy electron fluxes were recorded from a distance of about one planetary diameter above the surface of Mercury down to the closest approach made by Mariner 10: 470 miles (756 km).

The third experiment measures the strength and direction of magnetic fields by magnetometers mounted on a boom extending from the spacecraft. Independent measurements from two magnetometers located at different distances from the Mariner spacecraft allow scientists to eliminate the effects of the magnetic field of the spacecraft itself from the field being measured in the space through which the spacecraft passes.

In interplanetary space the magnetic field is typically about 6 gamma, compared with Earth's field at the equator amounting to 30,000 gamma.

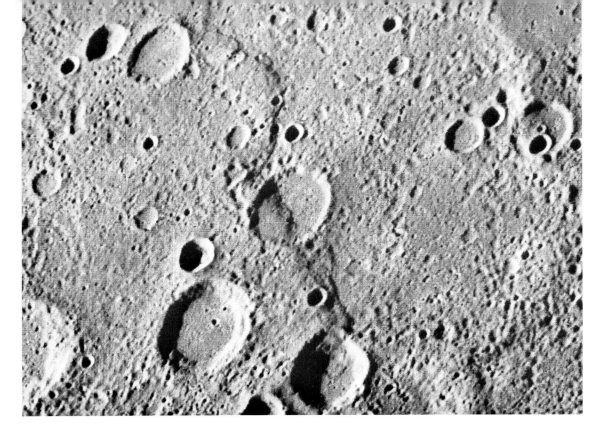

Figure 5.14 *SANTA MARIA SCARP*. This low-relief lobate scarp is located near 5°N latitude and 18°W longitude. It is named after Columbus's ship on the first voyage to America, 1492. The scarp conspicuously transects a 25-mile (40-km) diameter crater and is seemingly terminated by an only slightly smaller one at the bottom of the frame. A relation such as this suggests that scarps were forming during the last phase of heavy bombardment on Mercury.

Figure 5.15 *DISCOVERY SCARP*. A remarkably similar scarp to that shown in Figure 5.14 is located at 54°W longitude. The Discovery was one of Cook's ships on his last voyage to the Pacific, 1776–1779, when he was killed in the Hawaiian Islands. This scarp is about 340 miles (550 km) long and transects two craters of 34 and 21 miles (55 and 35 km) diameter. A peculiar shallow crater is perched on the crest of the scarp, just south of the larger, transected crater. There appears to be definite vertical and horizontal motion associated with the crater transectional relationship. Such relationships as well as the form and dimensions of the scarp suggest that these structures are associated with global crustal shortening on Mercury rather than the more familiar crustal extension that one sees in the lunar rilles and in many Martian features.

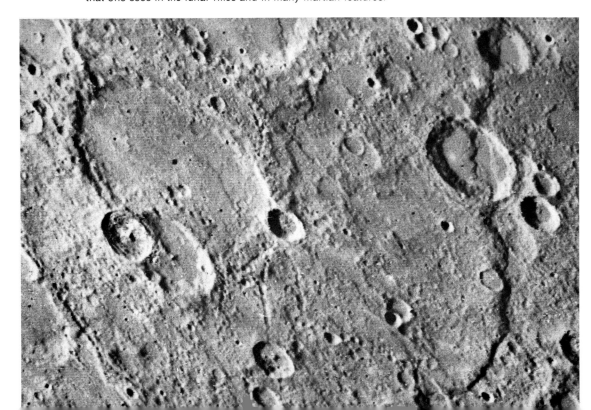

Figure 5.16 *VOSTOK SCARP, I.* This 80-mile (130-km) long scarp cuts two craters conspicuously in the left-hand portion of this frame. The northwestern rim of the lower, younger crater, 40 miles (65 km) in diameter, has apparently been offset about 6 miles (10 km) by the scarp. The lower (southern) portion of the scarp is eastward-facing, whereas the northern is westward-facing. The scarp is located about 38°S latitude and 17°W longitude, and the resolution of this picture is about 1 mile (1.7 km). Vostok was the name of the ship used by Admiral Bellingshausen in the Russian Antarctic Expedition of 1820–1821.

Figure 5.17 *VOSTOK SCARP, II.* This high-resolution (2000 feet, or 600 meters) view of the offset crater in Figure 5.16 clearly shows the west-facing portion of the Vostok Scarp. It has a slightly terraced appearance that has been speculatively attributed to individual thrust-fault planes. The spacecraft was only 10,930 miles (17,600 km) from the surface of Mercury at the time this frame was acquired. Note the intense pitting of the otherwise smooth-appearing crater, as well as several different levels of fill. The fuzzy band extending from left to right across the crater is due to a momentary lapse in data receipt during the mission. Other linear escarpments can be seen just northeastward of the crater, paralleling the Vostok Scarp and perhaps related to it.

As Mariner approaches Mercury, the magnetic field begins to increase, and it continues to do so until it reaches almost 100 gamma at closest approach. If this rate of increase continues to the surface, Mercury has a surface field of about 200 to 300 gamma. Such a field would be more than adequate to deflect the solar wind and produce the bow shock evidenced by the plasma and charged particle experiments.

Mercury, while looking like a twin of the Moon to the TV cameras, is beginning to look like a little Earth to particles and magnetic field instruments! Clearly Mercury possesses a magnetic field. But how does the field originate on such a slowly rotating planet? One possibility to be considered is that Mercury rotated much faster earlier in its history and generated a field by the dynamo effect—like the Earth. Perhaps this field could have been locked into the cooling planet to produce a permanent intrinsic field.

Another speculation is that perhaps a slowly rotating planet can maintain a magnetic field if it has the right kind of fluid, electrically conductive core. Mercury, with a density almost the same as the Earth's, is believed to have a large core of iron, whereas some other bodies that do not have magnetic fields, such as the Moon and Mars, do not have a large iron core. Alternatively, the field may be due to a complex mechanism associated with the interaction of the solar wind with the planet. The sweeping of interplanetary field lines past the planet might generate an electrical current flow in the planet, and this, in turn, would produce the observed magnetic field. Clearly another encounter with Mercury will be required to answer this "intrinsic vs induced" question.

In addition to electrical and magnetic measurements, precise observation of the radio signals from Mariner has been providing important data about the gravitational field of the planet; as the radio signals are interrupted by Mariner's passing behind Mercury as seen from Earth, new information is gained about the size of Mercury and about its hypothetical atmosphere.

Since the early telescopic observations of Mercury, astronomers have speculated about an atmosphere. Mariner 10 offers promise of clearing up the matter. The sensitivity of the radio science experiment permits it to detect any ionosphere exceeding 1000 electrons per cubic centimeter, a tiny fraction of the density of Earth's atmosphere. No evidence for either an atmosphere or an ionosphere of Mercury is found. Even a supersensitive ultraviolet-light detector aboard Mariner 10 discovers only a tenuous envelope of helium gas about the planet, probably derived from the solar wind.

The radio experiment also provides accurate data on the diameter of Mercury and on the planet's mass. These quantities are necessary to determine the density of the planet. Radio measurements of the precise path of Mariner past Mercury determine that the planet's mass is 1/6,023,600 that of the Sun. This figure agrees closely with the mass derived from measurements made by radar from Earth. The radius of 1,515 miles (2,439 km) is also in close agreement with earlier radar measurements. Mariner 10 confirms that the mean density of the planet is 5.44 times that of water—almost the same as for Earth.

Meanwhile, the imaging of Mercury is producing remarkable close-up pictures of the rugged surface. The major landforms are identified as basins, craters, scarps (steep slopes), ridges, and plains. Morphologically some of these features strongly resemble features on the Moon. For example, Mercury's surface is splattered with many bright ray craters similar to those found on the Moon (Fig. 5.18). The youngest craters have bright ray systems which are thought to be caused by spreading of debris across the surface. This debris darkens with age under the effect of micrometeorites, solar radiation, and the solar wind.

A large basin observed on the outward leg of the flyby is named Caloris Basin (Figs. 5.19 to 5.24). Its center is at 192°W and 30°N and it resembles a large impact basin on the Moon, with blankets of ejected material around it and concentric rings of mountains. Its floor is filled with smoother plains that appear ridged and fractured.

The plains materials present many of the features of the lunar maria and are dotted with craters to about the same degree (Figs. 5.25 to 5.30). Elsewhere the surface looks at first glance much like the highlands of the Moon, with overlapping craters forming rugged terrain.

Apparently, extensive flooding by rock materials grossly similar to those of lunar maria has occurred on Mercury. The large horizontal scale of these features implies a silicate composition with a density of about 3 for the entire outer region of the planet. The mean planetary density of 5.44 implies that much denser material must be present deep within the planet—very probably in the form of a large iron core, which is independently suggested by the strong interaction with the solar wind. Also, Mercury's irregular scarps, some as high as 2 miles (3 km), which trace out lobate courses (i.e., with scalloped edges) across the intercrater plains indicate that Mercury may have contracted around its core early in its history.

In the half of the planet observed by Mariner 10, Mercury, like the Moon, seems to exhibit a hemispherically nonuniform distribution of flooded basins. If this impression is valid, previous explanations of the nearside/farside differences on the Moon which involve processes peculiar to the presence of the Earth may require re-evaluation.

99

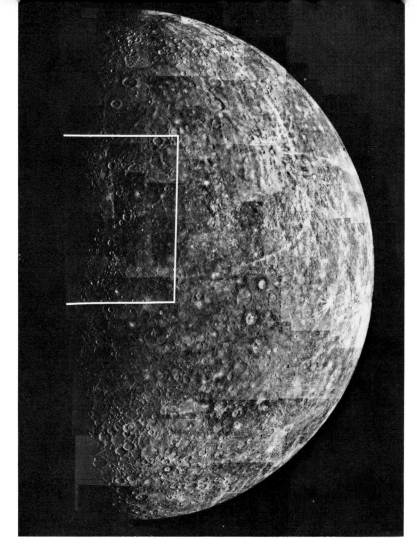

Figure 5.18 *THE OUTGOING HEMISPHERE OF MERCURY.* A view of Mercury as seen by Mariner 10 as it sped away from the planet on March 29, 1974. This picture is a mosaic of individual television frames acquired about two hours after encounter at a range of 37,300 miles (60,000 km). North is at the top, but now the terminator is at the left. The limb can be seen at the right, and the Sun's illumination is also from the right. The equator crosses from left to right about two-thirds of the way down the disc. The terminator is located at 190° W longitude, and the planet displays a gibbous (i.e., more than half-illuminated) disc in contrast to the appearance of the incoming fat-crescent view. The outgoing hemisphere is dominated by smooth plains rather than heavily cratered terrain and resembles portions of the lunar maria in general morphology. Half of a very large, multi-ringed basin named Caloris Basin appears near the center of the disc near the terminator. Its surrounding mountain ring is 800 miles (1,300 km) in diameter. The basin is shown in a specially processed mosaic in Figure 5.19.

Figure 5.19 *ZOOMING IN ON CALORIS, I.* This enhanced photomosaic shows details of the largest basin on Mercury seen by Mariner 10. It has been named the Caloris (meaning "hot") Basin because of its position near one of the subsolar points of Mercury at perihelion. An 800-mile (1,300-km) diameter ring of mountains up to 6,500 feet (2 km) in height defines the outer edge of the basin. Radial grooved structures are apparent beyond the mountains, especially in the northeastern portion of this view. The inner basin is filled by plains, presumably of volcanic origin, that are highly ridged and fractured. Other smooth, less deformed plains extend for hundreds of kilometers eastward. Parts of this basin are shown in four high-resolution frames in Figures 5.20–5.23.

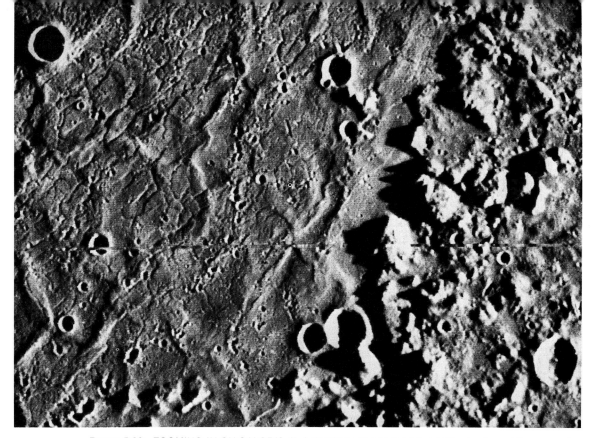

Figure 5.20 *ZOOMING IN ON CALORIS, II.* A detailed view of the edge of the basin and the rugged mountainous terrain surrounding it is shown in this high-resolution view, centered at about 20°N latitude and 181°W longitude. Resolution is about 0.6 miles (1 km). The area shown is slightly less than 175 miles (300 km) in width. A highly fractured and ridged terrain is displayed in this picture, as are chains of secondary impact craters laid down upon the material filling the Caloris Basin. The horizontal line through the lower center portion of the frame is caused by momentary loss of data.

Figure 5.21 *ZOOMING IN ON CALORIS, III.* This crater cluster, informally called ''Teddy Bear,'' is located on the east outer rim of the Caloris Basin at 32°N latitude, 173°W longitude. The largest crater is about 50 miles (80 km) in diameter. Resolution is about 0.6 miles (1 km). A dense field of secondary impact craters from the large crater is shown as it extends off to the east (right, in this view). There are several well-defined radial gouges made up of overlapping secondary impact craters immediately south of the large crater.

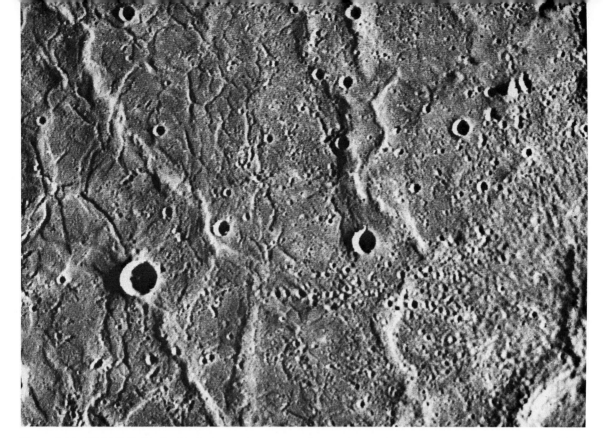

Figure 5.22 *ZOOMING IN ON CALORIS, IV.* This view of the plains of the Caloris Basin is immediately west (left) of the preceding frame. It shows the intense secondary-impact crater field from the ''Teddy Bear'' on the plains in the lower right-hand portion of the picture. Resolution, again, is about 0.6 miles (1 km), and the largest crater shown is about 9.3 miles (15 km) diameter. The presence of secondary craters on the Caloris plains demonstrates that the large craters of the ''Teddy Bear'' formed after the emplacement of the Caloris plains.

Figure 5.23 *ZOOMING IN ON CALORIS, V.* The wrinkled interior plains of Caloris shown here are located in the northern part of the basin near the terminator position at the time of Mariner 10's encounter. The frame is centered at 37°N latitude and 184°W longitude. Each of the two larger craters is about 30 miles (50 km) in diameter. Resolution is about 1.37 miles (2.2 km). The ridged and fractured character of the plains of Caloris shown in these frames does not have an exact counterpart on the Moon.

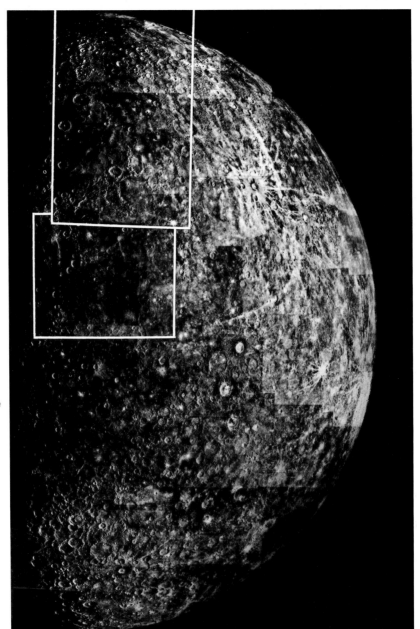

Figure 5.24 *THE PLAINS OF MERCURY, I.* Outside the Caloris Basin the smooth plains of Mercury often manifest a different, more Moon-like appearance than the surface within the basin. In the following series of photographs, two different areas outlined on this mosaic are shown at progressively higher resolution. The upper area is shown in Figures 5.25–5.27, and the lower area is shown afterward.

Figure 5.25 *THE PLAINS OF MERCURY, II.* Photomosaic of the northern plains of the outgoing hemisphere of Mercury. At the top (north), quite foreshortened near the limb, is a large basin [over 187 miles (300 km) in diameter], filled with smooth plains material. To the right and slightly below this basin is an extensive plains area, Borealis Planitia, which extends for over 500 miles (800 km). These and most of the other plains in this series are believed to be of volcanic origin, similar but not identical to that of the lunar maria. Another extensive smooth plains unit is in the center of the mosaic. The large craters just west (left) of the plains display well-developed ejecta blankets and secondary crater fields on top of the plains units, demonstrating that they are younger than the plains. The larger young crater is 50 miles (80 km) in diameter. At the bottom of the photograph an older, 138-mile (220-km) diameter crater has been completely filled by plains material that has also partially covered much of the ejecta deposits, indicating that the crater existed before the plains units. An area of plains material (lower left-hand side) is shown at high resolution in the following two figures.

104

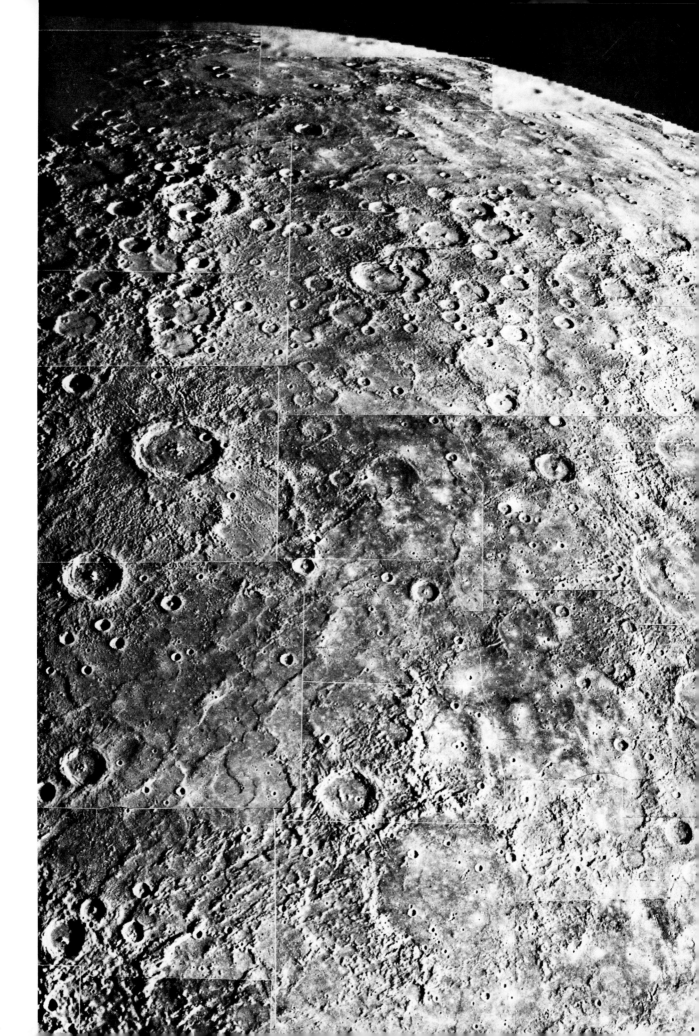

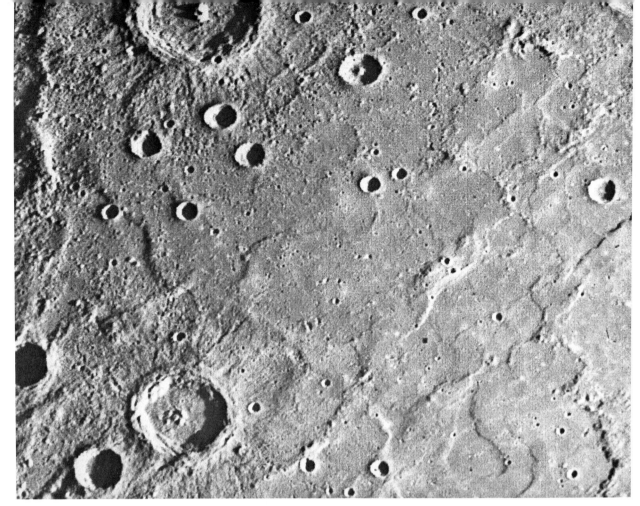

5.26
5.27

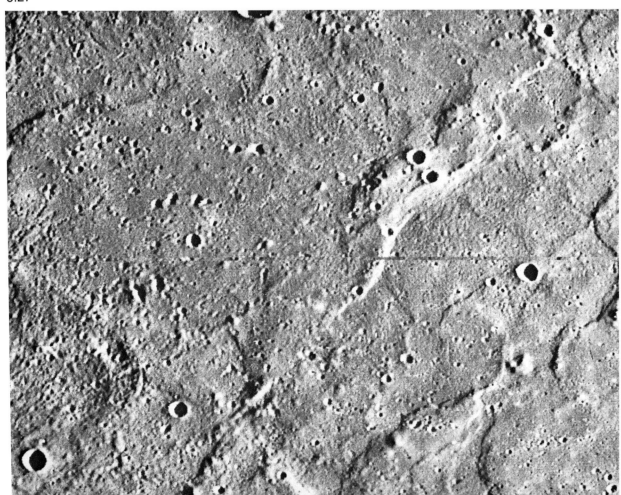

5.28

Figure 5.26   *THE PLAINS OF MERCURY, III.* The conspicuous, lighter-colored ridge running northeast-southwest (with twin craters) is characteristic of many of the plains of Mercury and is shown in even higher resolution in Figure 5.27. The young, fresh crater at the top of this picture is 56 miles (90 km) in diameter and exhibits a well-developed central peak, ejecta blanket, and secondary field radiating from it. The photo is centered at 50°N latitude, 170°W longitude, and has a resolution of 1.12 miles (1.8 km).

Figure 5.27   *THE PLAINS OF MERCURY, IV.* High-resolution view of the ridged character of the north central plains, centered at 50°N latitude, 170°W longitude. The picture covers an area about 150 miles (240 km) in width. Resolution is 0.56 miles (0.9 km). The alignment of many of the visible craters in chains and the formation of elongate structures are characteristic of low-velocity impacts. This suggests that the craters are more likely secondary craters, formed by ejecta from recent large impacts adjacent to the plains, rather than by asteroid or cometary objects impacting directly on the surface of Mercury.

Figure 5.28   *THE PLAINS OF MERCURY, V.* Displayed here are the edge of the Caloris Basin, the "Teddy Bear" crater, and the portion of Odin Planitia centered at 27°N, 160°W. (Odin is the Viking equivalent of the Roman word for the planet Mercury). In the center of the picture, just east of the Caloris mountain ridge, is an old, 125-mile (240-km) diameter crater that has been almost entirely filled by smooth plains material so that the crater is almost hidden, except for the remains of the rim. The arrow points to a small crater (just north of the rim of the large filled crater) on the portion of the Odin Planitia that is examined at progressively higher resolution in the succeeding two pictures. Resolution here is about 2.8 miles (4 km), and the diameter of the largest fresh crater in the "Teddy Bear" is about 50 miles (80 km). The picture is centered at 21°N latitude, 171°W longitude.

107

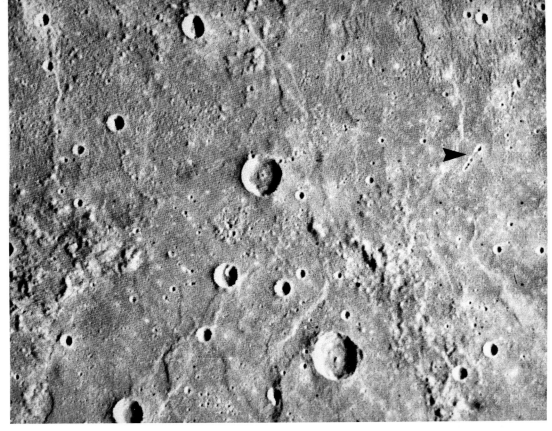

Figure 5.29 *THE PLAINS OF MERCURY, VI.* Moderate-resolution view of a portion of Odin Planitia centered at 27°N, 172°W. The largest of the fresh craters is about 9.3 miles (15 km) in diameter, and the resolution of this picture is about 0.6 miles (1 km). The north rim of the ancient filled basin is seen in greater detail in this view, as are the variety of scarps and ridges that characterize the smooth plains of Mercury. The arrow points to an alignment of fresh secondary-impact craters seen at high resolution in Figure 5.30.

March 30, 1974   With most of its mission to Mercury accomplished, Mariner 10 suddenly throws another of its tantrums. Late in the evening, just after the final close-in planetary pictures are completed, the spacecraft's power drain surges to 100 watts, greater than for any single piece of equipment in the spacecraft. Telemetered engineering data show a rapid climb of temperature. The instrument bay is heating up!

The spacecraft is in serious trouble, and scientists are pushing for a heroic, terminal frenzy of data collection before the whole mission falls apart. Jim Dunne, the project scientist, rushes into Giberson's office and pounds his desk to emphasize his contention that the experimenters' demands be ignored. This post-encounter power problem in the spacecraft brings several otherwise unflappable managers close to the point of desperation. Eventually calm is restored, but the period of desperation and near hysteria takes its toll, contributing significantly to problems concerning a satellite of Mercury aired at the April 1 press conference.

At 2:00 A.M. the next day, the temperature is still rising. Project management decides that the TV cameras and some other power-consuming instruments must be switched off. Then the rise in temperature halts.

Figure 5.30 *THE PLAINS OF MERCURY, VII.* One of the highest-resolution pictures acquired by Mariner 10 of the plains of Mercury displays features as small as 500 feet (150 meters) in dimension. At this high resolution, the surface is seen to be covered by both small primary craters and many secondary-impact craters. The chain of secondary craters at the bottom center is about 12 miles (20 km) long. The lobate scarp facing eastward (right) in the left portion of the picture is part of a ridge about 30 miles (50 km) wide. Photo is centered at 29.4°N, 159.4°W and was taken from a range of only 4200 miles (6800 km). This particular view is almost indistinguishable from scenes of similar resolution and illumination that was acquired by Ranger spacecraft when they first explored the mare regions of the Moon in 1965.

But the puzzle remains. What could be causing the use of 75 to 100 watts of power? This is 20 percent of the maximum capability of the spacecraft's power system. Only the main radio transmitter on the spacecraft consumes so much power. A short circuit of this magnitude could be a prelude to disaster for the spacecraft.

"Ed, I don't think this project can handle another crisis—everyone is just too exhausted. . . . And what's this rumor about the ultraviolet spectrometer team thinking they've discovered a satellite of Mercury?"

"Well, Bruce, apparently there is something in their data they can't explain."

Does Mercury have a satellite? Although Leverrier speculated in 1859 that a hypothetical planet, Vulcan, revolving around the Sun at a distance of 13,000,000 miles (21,000,000 km) caused a peculiar motion of the perihelion of Mercury's orbit (now believed due to relativity effects), no one appears to have postulated that Mercury has a satellite—not seriously, at least, until the Mariner 10 mission. There is a plan developed over several years for Mariner 10 to search with television pictures around Mercury for any captured asteroidal-sized bodies too small to have been observed from Earth. The strategy is to wait until all the close-up photography of the planet's surface has been acquired before risking time and effort to chase hypothetical satellites.

But the Mariner 10 press conference on Sunday, April 1, sparks an unprecedented speculation about a satellite of Mercury.

An unexplained ultraviolet emission had been detected several days before encounter; it showed up off the limb of Mercury during eight scans of the ultraviolet spectrometer. The source of the radiation moves away from the planet at a regular rate, and the spectrum is very different from that of Mercury. The scientists involved discuss with the press the possibility that the instrument has detected ultraviolet sunlight reflecting off the surface of an unknown satellite of Mercury (a satellite which somehow would have had to remain undetected in previous photography from both Earth and Mariner 10).

"Ed, have they checked out the possibility of the interference in the UV from one of those little particles traveling along with the spacecraft—the ones that have caused all the trouble with the star tracker?"

Reporters rush to telephones. The following day's newspapers carry headlines: Mercury May Have Moon, Data From Mariner Indicate," "Mariner 10's Data Indicate . . . Possible Moon," "Mariner 10 Reveals Mercury Has a Moon," "Possible Moon of Mercury Detected."

"Bad news, Bruce. The navigation engineers have just proved that the unexplained UV signal detected by the spectrometer corresponds exactly with the position of a 5th magnitude early-type star. They never saw any satellite—just a star."

"Mercury's Moon" is not a satellite but a distant star, catalogued as 31 Crater, appearing to move relative to Mercury because of the spacecraft's motion. The bubble of Mercury's satellite bursts on April 1 and proves itself a stellar April Fool's joke: just a gleam in Mariner's eye! Whether Mercury really does have any tiny satellites will have to wait for the planned TV search later. Despite the new spacecraft problems, the TV satellite search is carried out—and confirms that Mercury has no satellites as large as about 3 miles (5 km) in diameter, if indeed it has any at all.

Encounter is now completed, but with the unresolved spacecraft problems there seems little hope that Mariner 10 can survive to return to Mercury again to complete the photo survey of the illuminated hemisphere of the planet. It has been an extraordinary encounter, but it leaves everyone drained emotionally and physically.

110

○   ○   ○

Before Mariner leaves Mercury there is an important flashback: On July 31, 1964, Dr. Gerard P. Kuiper, standing on the stage at JPL's Von Karman Auditorium, presented the preliminary results of the Ranger 7 mission, the first successful American mission to another world. Ranger 7 had just returned the first close-up pictures of the Moon. Dr. Kuiper had said, "This is a great day for science and a great day for the United States." Dr. Kuiper was the principal investigator of the Ranger television team, and that mission initiated the exploration of the terrestrial planets that has led to the Moon, Mars, Venus, and now Mercury being inspected by spacecraft from planet Earth. Mariner 10 completed this exploration.

○   ○   ○

Professor Kuiper, pioneer and planetary astronomer, was also a member of the Mariner 10 television team. Unfortunately he died on December 21, 1973, while the spacecraft was still en route to Venus and Mercury. It is fitting that the center of the first feature observed on Mercury by Mariner 10, a conspicuous bright-rayed crater, should be named in his honor. N. W. "Bill" Cunningham, Mariner Venus/Mercury Program Manager of NASA Headquarters, in a rare moment of public emotion proposed that this crater be named after Dr. Kuiper, and NASA made the recommendation to the appropriate commission of the International Astronomical Union.

Crater Kuiper is centered at 11°S latitude and 30°W longitude. On all the pictures obtained by Mariner 10, longitude is defined by passing the 20° meridian through the center of a small crater tentatively named Hun Kal. The Mariner zero of longitude differs by about half a degree from that accepted by the International Astronomical Union prior to the flight.

April 3, 1974   ED REINECKE, LT. GOVERNOR OF CALIFORNIA, CONVICTED OF PERJURY RELATED TO ITT DEAL TO SUBSIDIZE 1972 REPUBLICAN NATIONAL CONVENTION

April 5, 1974   COALITION GOVERNMENT OF RIGHTISTS, NEUTRALISTS, AND PRO-COMMUNIST PATHET LAO ESTABLISHED IN LAOS

April 5, 1974   DWIGHT CHAPIN, NIXON'S FORMER APPOINTMENTS SECRETARY, CONVICTED OF PERJURY RELATED TO POLITICAL "DIRTY TRICKS" IN 1972 PRESIDENTIAL CAMPAIGN

April 10, 1974   Mariner 10, in its cruise mode and almost out of stabilization gas, enters its extended mission as it starts on its long orbit around the Sun to try for a seemingly impossible second rendezvous with Mercury six months later.

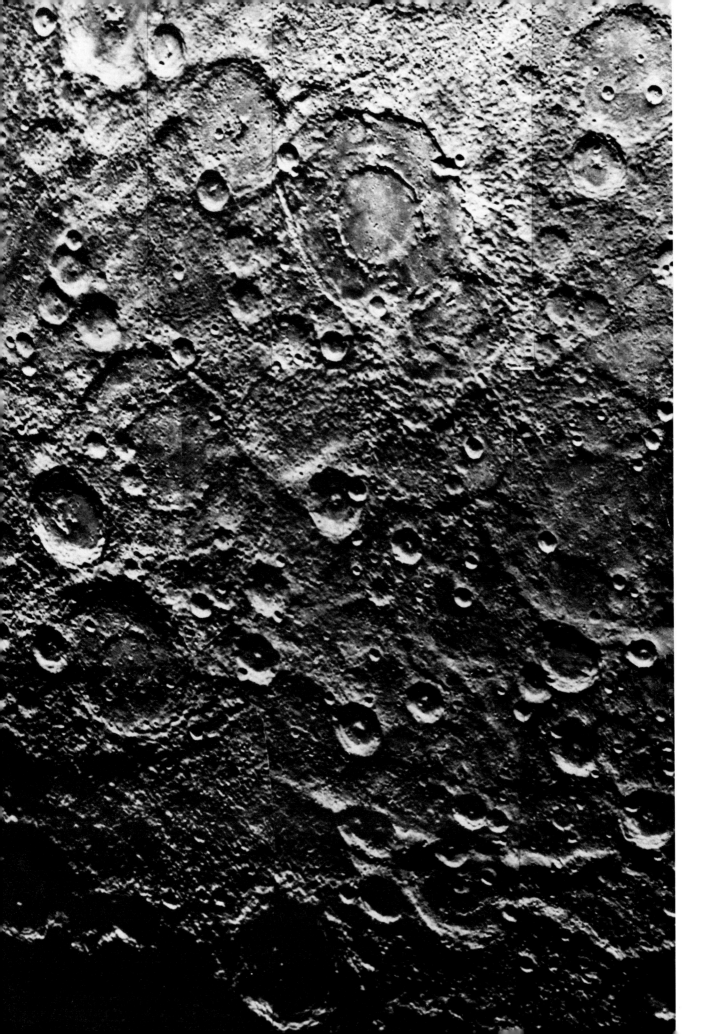

6

Return to Mercury

"Dr. Murray . . . can the spacecraft be made to come back?"
"Come back, Dr. Colombo?"
"Yes. The spacecraft could return to Mercury."
"Are you sure?"
"Why don't you check?" —Caltech, February, 1970

Once the excitement, and exhaustion, of Mariner 10's first encounter with Mercury subsides, JPL and NASA begin to consider the future. Of course, Mariner 10 had been upgraded before launch to permit in theory a second and even a third encounter with Mercury at successive six-month intervals. But the many difficulties encountered after launch had made first the Venus encounter and then the initial Mercury encounter very uncertain. Hence, future possibilities had been virtually written off. But the spacecraft is still "alive." In principle, a second encounter is a practical objective. As Vic Clarke is quick to point out, extremely careful management of the attitude stabilization gas and of the rocket propellant to carry out trajectory corrections could permit even a third passage by the planet. After all the troubles, however, can one seriously hope for yet additional achievements, for a successful extended mission?

Inexorably, while scientists and engineers and management review probabilities, Mariner 10 moves along its orbit around the Sun.

May 2, 1974   FORMER VICE PRESIDENT SPIRO T. AGNEW DISBARRED BY
MARYLAND COURT OF APPEALS

Unless a trajectory correction maneuver is carried out, Mariner 10's path around the Sun will bring it back for a second encounter with Mercury at a range of 500,000 miles (805,000 km). This "miss distance" is much too great for imaging of the sunlit surface or acquisition of any other useful observations. Nor would it lead to a third encounter. So Mariner 10 must be maneuvered into a slightly different path to bring it closer to Mercury and also to allow a third encounter. But this cannot be carried out by a "sun-line" maneuver. The risk of reorientation and possible attitude-control oscillation is now unavoidable. Moreover, there is much controversy about the targeting for the second encounter. The discovery of a substantial magnetic field of Mercury at the first encounter was unexpected. Many questions are raised about this field; is it intrinsic to Mercury or the result of some not yet understood characteristic of the interaction of the solar wind with the planet? Because another mission to Mercury is not likely for a considerable time, some scientists naturally want a second approach of Mariner 10 that will permit further evaluation of the magnetic field and the interaction of Mercury with the solar wind. However, such an approach is not conducive to photo-coverage of regions that had not been viewed at the first encounter because of the nightside pass. The team members vividly remember the Mariner 9 orbiter flight about Mars, when totally new features were discovered in regions of the planet not photographed by the earlier flybys of Mariners 4, 6, and 7.

Jim Dunne, the project scientist, argues strongly for a brightside passage at the second encounter, basing his argument on the Mariner 9 experience. Moreover, he points out that a distant passage on the day hemisphere would increase the probability of achieving a third encounter with Mercury with the limited gas available within the spacecraft. On the other hand, making the second encounter a darkside pass to check on the magnetic field would not provide new photo-coverage and would have a high probability of not permitting a third return to Mercury.

Still, everyone knows that a third encounter really is a long shot and that Mariner may never be able to find out if the magnetic field is intrinsic to Mercury unless the second flyby is made over the dark side.

After many hours of discussion, NASA Headquarters directs that the second encounter will aim for a dayside pass to give good photo-coverage of the sunlit hemisphere of Mercury. The next TCM is planned. It requires a relatively large change in the velocity of the spacecraft and as a consequence the rocket engine will have to be fired twice to avoid overheating.

May 9, 1974
12:44 P.M. PDT

The spacecraft begins its first roll turn followed immediately by a pitch turn, completed without incident. Ignited at 1:05 P.M. PDT, the propulsion engine burns for 195 seconds. All goes well. The spacecraft is pitched and rolled back to its cruise position and the celestial reference is acquired again. At 12:44 P.M. PDT on May 10, the gyros are again used to roll the spacecraft through 178 degrees followed by the necessary pitch turn. The rocket engine is fired at 1:06 P.M. PDT and burns for 139 seconds. The two maneuvers have corrected the path of the spacecraft to within 1 percent of the velocity change required. The spacecraft is pitched and rolled back to its cruise orientation and the star reference is reacquired.

During this maneuver the use of attitude control gas is entirely normal; no oscillations occur. The remaining gas supply now seems adequate for the return to Mercury. And if all goes well, there might be barely enough gas left for a third encounter.

May 17, 1974 SIX SYMBIONESE LIBERATION ARMY (SLA) MEMBERS KILLED IN LOS ANGELES SIEGE

May 18, 1974 INDIA EXPLODES NUCLEAR DEVICE

June 3, 1974 PRESIDENTIAL STAFFER, CHARLES W. COLSON, PLEADS GUILTY TO ATTEMPTING TO OBSTRUCT JUSTICE IN MATTER OF ELLSBERG TRIAL

June 6, 1974 Mariner 10 is some 148 million miles (238 million km) from Earth, traveling on the far side of the Sun. Communications are temporarily interrupted.

When Mariner 10 emerges from behind the Sun, and the trajectory is further refined by observation of the radio signals, it is clear that the correcting maneuver has been very successful and that a third encounter with Mercury is possible. To make the third encounter actually take place, however, the path of the spacecraft will have to be corrected slightly once again, in a fifth TCM about the beginning of July.

The fifth maneuver does not require a large velocity change, only 10.85 feet per second (3.31 meters/sec). But it is a tricky maneuver; to perform it the spacecraft must break communication with Earth and rely entirely on its internal memory to make the maneuver and turn the spacecraft back so that it can again communicate with the controllers at Pasadena. The risk must be accepted. A strategy of trying for two more encounters with Mercury had been decided upon. The possibility of the third encounter, although rather remote, has to be kept open by making this maneuver.

The sequence of commands necessary for the maneuver is transmitted to the spacecraft. A few minutes before 1:00 P.M. PDT on July 2, the roll commences and oscillograph pens tracing telemetered data from the spacecraft at the control center stop wiggling. The roll of the spacecraft has turned the antenna away from Earth, and signals can no longer be received. This is a time of considerable anxiety because there is no way

115

of knowing whether the roll oscillations are occurring and whether the precious maneuvering gas is being wasted. If it is, there will be no further encounters with Mercury because the spacecraft simply cannot survive if not oriented properly relative to the Sun.

The rocket engine fires at 1:09 P.M. for 18.8 seconds to provide the velocity change. Changes in the speed of the spacecraft relative to Earth will cause a change in the frequency of the spacecraft's radio signals reaching Earth. This is called a Doppler shift. The first indication that all is well comes from tracking data. The Doppler shift in the faint signal received at the Deep Space Network Goldstone Station is about the right amount. A short while later as the internally stored commands of the spacecraft execute the pitch and roll to bring back full communications, the pens on the recorder start to wiggle again, engineering data come flooding back to Earth, and the spacecraft engineers know that everything is all right. No gas has been wasted. Mariner 10 is heading successfully for a second encounter, with good possibilities of a third.

The brightside pass of the second encounter is targeted for a closest approach of 30,000 miles (48,000 km) over the southern hemisphere. Exactly the same hemisphere of Mercury is illuminated at the first and second passes because during the six months in which the spacecraft made one orbit of the Sun, Mercury made two orbits and rotated on its axis three times. The new coverage will provide photographs of the areas between the two sets of pictures obtained on the incoming and outgoing paths of Mercury I.

August 8, 1974   RICHARD NIXON RESIGNS AS PRESIDENT OF THE UNITED STATES

August, 1974   Bad news from Mariner 10. The tape recorder on the spacecraft fails completely. It is now impossible to store pictures on board and transmit them at a low communication rate later. If all the pictures that Mariner 10's two cameras are capable of taking during the second flyby are to be transmitted to Earth, they must be sent over the radio links as rapidly as they are taken, in "real time." This requires the highest communication-rate capability of 117.6 kilobits per second, using the 210-foot (64-meter) antenna at Goldstone and its ultra-low-noise receiving cone. The high-rate communication system is no longer just an extra achievement—it is essential for a useful second encounter. In addition, two smaller antennas are connected in an array with the big antenna at Goldstone, using microwave links to make, in effect, an even larger antenna.

In the meantime, Mariner 10 approaches Mercury. Its controllers continue to use a form of solar sailing to conserve maneuvering gas.

September 8, 1974   PRESIDENT FORD UNCONDITIONALLY PARDONS RICHARD NIXON; NIXON ACCEPTS PARDON

| September 9, 1974 | The spacecraft's memory is updated with a series of commands for it to go through its encounter sequence automatically. |
| | |

On September 20, television transmission starts in earnest. Twelve six-picture mosaics of Mercury are transmitted with a resolution of 12 miles (19.3 km).

Closest approach occurs on September 21 at 1:59 P.M. PDT, as Mariner 10 passes over the sunlit hemisphere. The range to Earth is now 105 million miles (169 million km), and great ingenuity is needed to get the millions of bits of imaging data back to Earth during the short time of the flyby. The novel three-antenna linkage at the Goldstone tracking station is successful, and all the high-resolution pictures obtained by Mariner 10 are returned safely to Earth. About 2,000 photographs are obtained, many sequences showing similar areas of Mercury's surface but from much more favorable viewing angles, thereby providing excellent new scientific data about the surface of the planet.

Preliminary interpretations from the first encounter are verified and extended (see Figs. 6.1, 6.2, and later figures showing high-resolution views). Mariner 10's string of accomplishments is extended. But can the last major objective—the third encounter and an additional look at the magnetic field—be achieved?

September 22, 1974 The spacecraft is ordered back to its cruise mode, TV cameras are turned off, and the platform on which the cameras are mounted is commanded to return to its stowed position. The spacecraft virtually hibernates. Project engineers are happy to see that Mariner 10 still has enough rocket propellant and maneuvering gas for two more maneuvers, if all goes well.

September 23, 1974 FORMER PRESIDENT RICHARD NIXON ENTERS HOSPITAL WITH PHLEBITIS

To attain the third encounter is a major technical challenge. The space-craft was not designed originally for such a long voyage. The roll oscillations have used most of the maneuvering gas reserves, the internal tape recorder is not functioning, and there are six more months to go. Furthermore, many key personnel at JPL and Boeing are being transferred to new jobs. Engineers and scientists still connected with the project must devise even more effective techniques to minimize the use of maneuvering gas and yet keep the spacecraft under control in its long voyage around the Sun.

October 6, 1974 Just two weeks after the second encounter with Mercury the spacecraft is searching for Canopus and roll-axis oscillation occurs, spilling 0.2 pounds (0.09 kg) of nitrogen gas into space. Only 0.6 pounds (0.27 kg) are now left for the six-month cruise to Mercury III. The spacecraft is now allowed to roll freely, and no attempt is made to keep it precisely

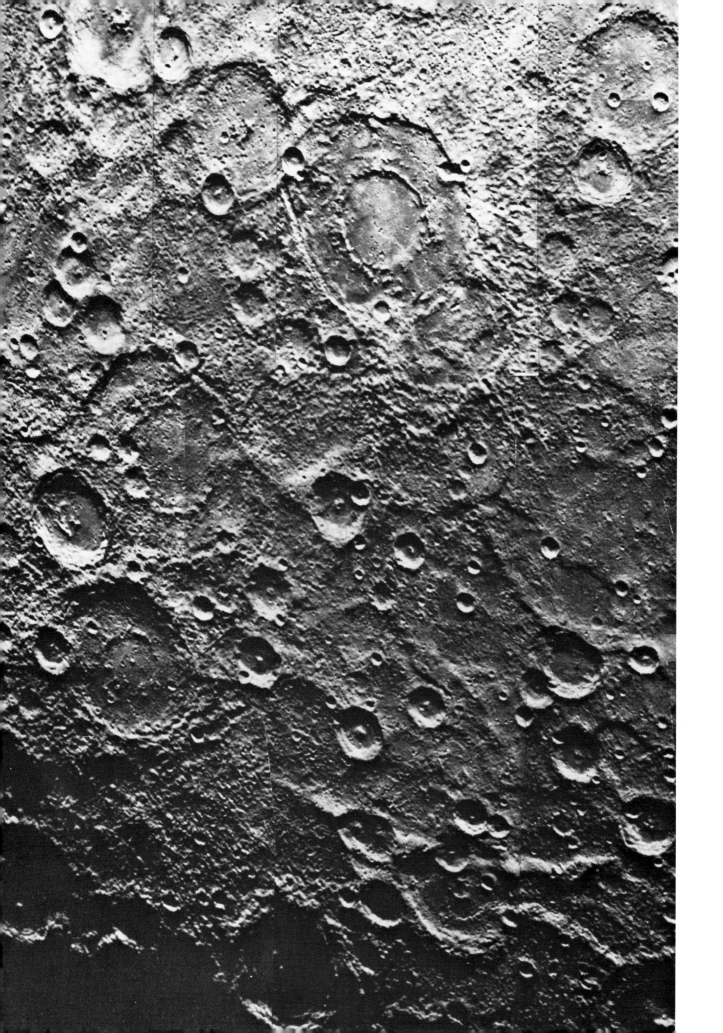

oriented. Solar pressure is used to keep the roll rate within limits. The position of the spacecraft at any instant is computed on the basis of the constant roll rate and checked against the way in which the pattern of radiation received from the antenna on the spacecraft varies at Earth, and how stars are observed to cross the field of view of the star sensor. The telemetry channel that tells the brightness of the stars in the field failed in the electrical disaster just after the first encounter. But controllers know the limit of the sensor and know which of only three acquirable stars could be passing across its field of view due to the roll of the spacecraft. The only continuous indicator of roll position is the pattern of radio radiation from the low-gain antenna, which unfortunately had not been mapped in detail before launch. This provided another "learn as you go" aspect of the Mariner 10 operation.

A new problem is emerging: Mariner 10's low priority for station coverage relative to other spacecraft—Pioneer 10 beyond Jupiter, Pioneer 11 on its way to Jupiter, and the joint U.S./Germany probe, Helios, heading toward its closest approach to the Sun. Yet, when Mariner 10 is transmitting over the low-gain antenna for its communications with Earth (as is necessary in the rolling-free mode), the big antennas at the tracking stations of the Deep Space Network are required for most of the period from initiation of solar sailing on October 6 to the third encounter with Mercury the following March.

The hibernating Mariner continues silently to orbit the Sun. Twice it becomes necessary to correct the path of the spacecraft, each maneuver being aimed at bringing the spacecraft's anticipated arrival and distance from Mercury closer to the desired values. Suddenly, after several weeks of observation following the second such maneuver, controllers find that Mariner is approaching too close to the planet. It might crash onto Mercury's surface! The desired flight path has been chosen to be quite close to Mercury so that the magnetic field can be measured most accurately, but Mariner 10 must not crash onto the planet.

Fortunately the maneuver strategy had been very carefully worked out in advance. After TCM-6, Dunne, who has in effect been running the project since the second Mercury encounter, decided to delay TCM-7 in

Figure 6.1 *THE SOUTH POLE OF MERCURY.* This photomosaic shows a portion of the heavily cratered terrain surrounding the south pole of Mercury as observed by Mariner 10 on its second pass by Mercury on September 16, 1974. The pole is located in the center of the large crater near the bottom center of the page, almost completely shadowed. At the top is a well-preserved 140-mile (225-km) double-ring basin. It is located at approximately latitude 70°S. The 105th meridian runs from the south pole almost vertically to the top of the photo. The Sun is coming from the top (north), and the viewer is looking at the terminator region. The resolution is between 0.9 and 1.2 miles (1.5–2 km) in the pictures used for the mosaic, which were acquired from a distance of 35,000 miles (56,000 km).

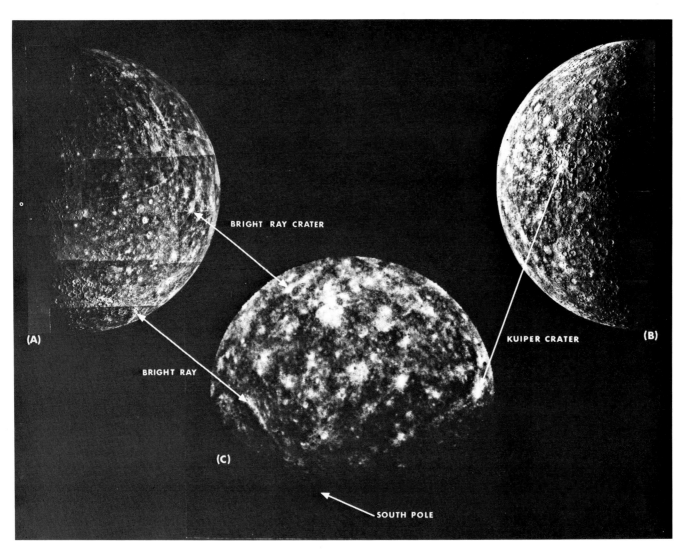

Figure 6.2 *TYING THE HEMISPHERE TOGETHER.* This diagram relates the incoming hemisphere as seen by Mariner 10 (*B*) with the outgoing hemisphere seen on the first pass (*A*) to the south polar view acquired by Mariner 10 in its second pass (*C*). Conspicuous bright-rayed craters present in various combinations of the views are indicated by arrows. The morning terminator is on the left, the evening on the right. North is toward the top of all the illustrations, and illumination is from the center of the diagram. The diameter of Mercury is 3,030 miles (4,878 km).

Figure 6.3 *AN ENHANCED VIEW OF THE SOUTHERN HEMISPHERE OF MERCURY.* To make maximum use of the hundreds of individual high-resolution frames acquired by Mariner 10 as it sped over the south pole of Mercury, a selected set was individually processed and projected as if seen from a common direction. That direction passes through an imaginary point on the surface at 105°W longitude and 60°S latitude. The south pole can be again recognized in the large crater on the terminator at the center. Several of the bright-rayed craters seen in the low-resolution view on Figure 6.2 can be identified in more detail in this version, including crater Kuiper. The upper horizon as seen in this view is approximately the equator.

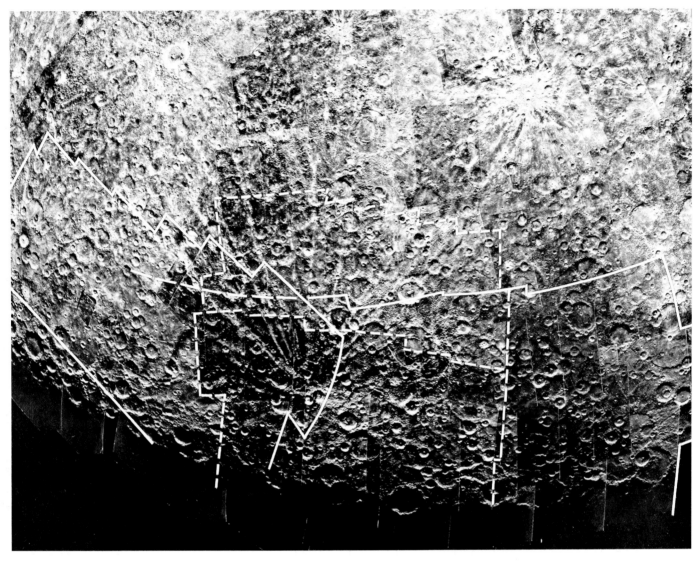

Figure 6.4 *OUTLINES OF HIGH-RESOLUTION VIEWS.* In this enlarged portion of the preceding photomosaic, the crater containing the south pole can be recognized in the lower center portion of the picture. High-resolution mosaics are shown in Figures 6.5–6.7 and are outlined in white here. Distance from the south pole to the top of the photomosaic is approximately 1,240 miles (2,000 km). The large, bright-rayed crater in the upper left is at 120°W longitude and 26°S latitude. The conspicuous bright-rayed crater with central peak in the center of the extreme left portion of this mosaic is seen also at the top of Figure 6.5 and is located at 164°W longitude and 31°S latitude.

Figure 6.5 *A CLOSER VIEW, I.* This is the high-resolution view of the area outlined in white at the left of Figure 6.4. It shows a heavily cratered terrain interspersed with smooth plains of differing ages, along with two conspicuous bright-rayed craters. The south pole is just off the field of view at the bottom. The center of the photomosaic is approximately along the 150° meridian. Average resolution is about 1.2 miles (2 km). Numerous scarps are seen, up to several hundred kilometers long, which transect and distort large craters. In the lower left-center of this view is a prominent scarp more than 185 miles (300 km) long, which has been named the Hero scarp in honor of the ship used by the American explorer Nathaniel Palmer in his 1820–1821 exploration of the regions of the Antarctic coast. The bright-rayed crater with the dark rim and central peak (upper left-hand side) is 42 miles (67 km) in diameter and located at 164°W longitude and 31°S latitude. The dark-rimmed and generally bright-rayed character is similar to the crater Tycho on Earth's Moon. A large, badly degraded old basin 217 miles (350 km) in diameter, located at left center, is filled with smooth plains and in turn has been cratered extensively at later times.

order to get a closer flyby of Mercury and possibly avoid the need for an eighth maneuver. Comments Dunne:

"The thing which allowed us to adopt this strategy was that we had a sun-line 'backaway' option on March 7 should the seventh maneuver, TCM-7, push us in too close to the planet. As it turned out, we had to exercise the backaway.

"There always was some uncertainty, even after the TCM-8 backaway. The last scare of the mission occurred a couple of hours before encounter when the real-time trajectory readout started to climb rapidly, indicating an impending impact on Mercury! Although none of us was inclined to believe the real-time program, it certainly ruined my lunch! And it made clock-watchers of us all for the period of periapsis [closest approach to Mercury]."

"A few minutes before closest approach, a newsguy asked me what we were all watching for, and I replied, 'For the lights [meaning telemetry] to go out.' He said, confused, 'Which lights should I watch?' "

March 7, 1975
2:00 A.M. PDT

The roll of the spacecraft is stopped. The gyros are used to orient the spacecraft in the correct direction for its rocket engine to thrust and provide the correcting velocity. Fortunately there are no untoward incidents of roll oscillation. The flight directors have determined precisely how the spacecraft can be maneuvered to avoid it. But the roll oscillation–avoidance techniques place many constraints on how the spacecraft can be operated. At this TCM the rocket engine blasts for a mere 3 seconds, thrusting the spacecraft through a small velocity change of just over 1 foot per second (0.3 meters/sec). The aiming point at Mercury is lifted another 100 miles (160 km) above the surface to lessen the chance of a collision. Closest encounter is now scheduled for 3:39 P.M. PDT, on March 16, 1975, at an altitude of 125 miles (200 km) at about 70°N latitude—the closest flyby of any planet.

Imaging is planned to concentrate on high-resolution mosaics of areas of special interest seen in pictures from the earlier flybys. But highest priority is for the particles and fields experiments to try to determine whether the magnetic field of Mercury observed during the first encounter is somehow connected with the solar wind or, instead, is intrinsic to the planet itself.

The third encounter nears. Mariner 10 is rolling slowly through space, making one revolution every 60 hours. Controllers are trying to decide how soon this top-like revolution should be stopped and the spacecraft oriented for its flyby. If they put Mariner under control too soon, an anomaly could use up all the maneuvering gas before encounter and cause the spacecraft to fly by uselessly, unable to send its data to Earth or direct its instruments at the planet. If they are too late, there might not be time to assure a lock on Canopus and to establish correct orientation for the flyby. Again risks must be taken.

124

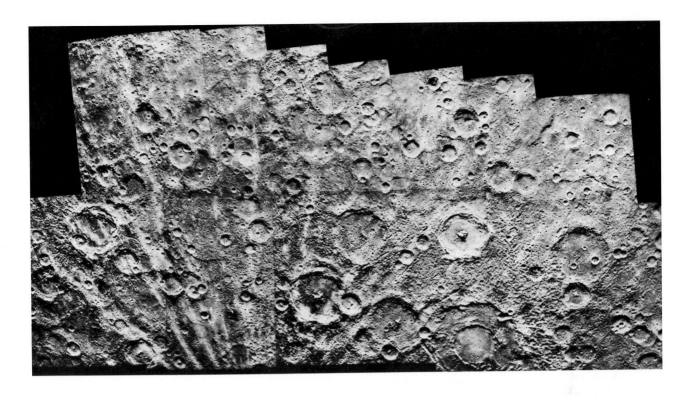

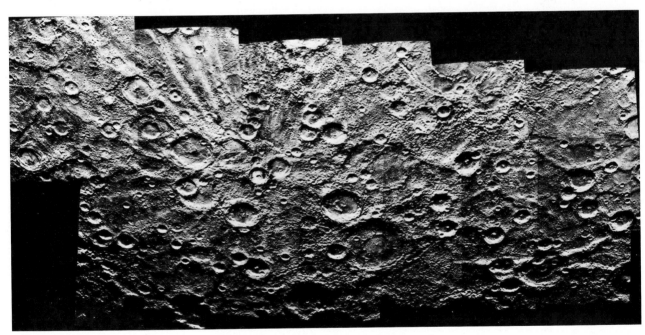

Figure 6.6 *A CLOSER VIEW, II.* These two, slightly overlapping photomosaics provide a
high-resolution view of the central portion of the area outlined in Figure 6.4. The family
of rays radiating from a bright crater approximately 43 miles (70 km) in diameter in the
lower central region is the most conspicuous feature here. Rays from this crater extend
over 620 miles (1,000 km), even outside the field of view. The 140-mile (225-km) double-
ring basin seen in Figure 6.1 is again shown at the right central portion of this mosaic.
The crater enclosing the south pole is right at the terminator, directly below the double-
ringed basin. Average resolution is about 1 mile (1.6 km).

125

The low-gain antenna patterns are observed and from them it is deduced that the first good chance to acquire Canopus before encounter will be on March 12. Unfortunately, at that time the spacecraft will be between Deep Space Network stations in Spain and Australia in the midst of a three-hour tracking gap. A signal is sent to the spacecraft in order to attempt to change the roll rate and delay the passage of Canopus through the star tracker's field until the spacecraft is being tracked by the DSN station in Australia. This attempt is unsuccessful.

To complicate matters, many hours of use of the big antennas have been allocated to other spacecraft, especially the interplanetary probe Helios which is nearing its closest approach to the Sun, the most important point in its mission. The Helios project agrees to give up some of this valuable time for the Mariner emergency.

March 13, 1975   A day of serious problems that almost led to failure of the third encounter. Jim Dunne recounts the dramatic experiences of these few days before the third encounter:

"Late in the day of March 13, we passed through Canopus. All that remained was the simple task of backing up the spacecraft in its roll, passing Canopus through the star sensor once again, and then driving the spacecraft forward in roll again to an easy acquisition of the guide star. I went home relieved.

"Somewhere between 2:00 and 3:00 a.m. on March 14, I was called at home with the news: 'We've rolled into the big null.' The spacecraft had not rolled far enough back when the roll had been reversed and there was no telemetry indication of when the star passed through the tracker when the roll was again made in the positive direction. Now the signal strength from Mariner 10 plummeted as the low-gain antenna's low radiation point faced Earth. Commands were sent to back the spacecraft away from the null in the low-gain antenna signal pattern but they were not acknowledged because we had lost telemetry.

"We were in trouble.

"The spacecraft could have been rolling in either of two directions or, conceivably, stopped dead. We didn't have DSS 14 [the big antenna at Goldstone station] available for Mariner 10 that day, and DSS 63 set [the big antenna at Robledo, Spain was turned away from the spacecraft by Earth's rotation] before telemetry was recovered."

Later that day Dunne had to participate in a pre-encounter press conference in the Von Karman Auditorium at the Jet Propulsion Laboratory, where the science reporters were hearing the bad news that Mariner 10 could very well fly past Mercury in silence, completely out of touch with the Earth and so disoriented that its instruments could not probe the innermost planet. Dunne continued:

"At the end of the press conference I received the good news that station 12 [through Echo, the smaller antenna at Goldstone] had received the

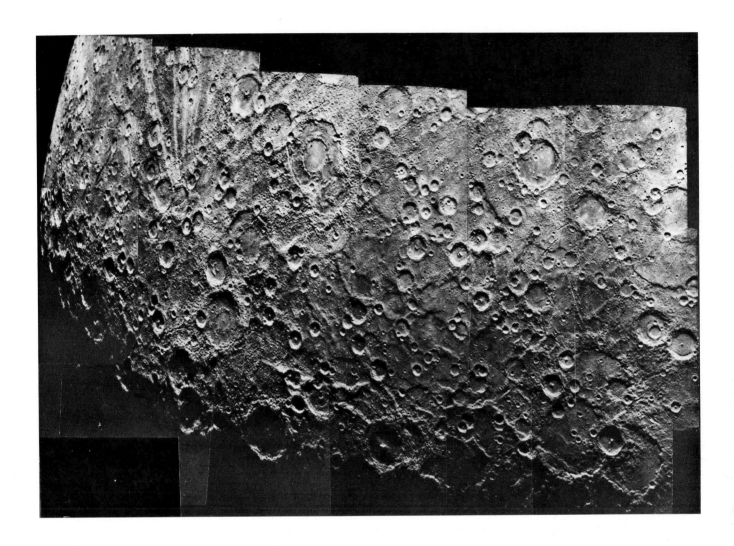

Figure 6.7 *A CLOSER VIEW, III.* This third view, moving from west to east across the south polar regions of the planet, shows the crater enclosing the south pole clearly displayed on the terminator just to the left of the lower center of the frame. The bright-rayed crater and double-ringed basin seen in Figure 6.6 can also be recognized in the left-center portions of this view. At the extreme right is Discovery scarp, one of the prominent features seen in the first Mercury pass. (The sequence in which these three pictures are printed here is the opposite from that in which they were acquired. The spacecraft approached Mercury from the right-hand side as seen in these views and moved to the left-hand side, connecting the two hemispheres first seen by Mariner 10 on March 29, 1974.) To the left of Discovery scarp is another series of scarps, 250–300 miles (400–500 km) long, which has been named the Fran scarp, after the Norwegian ship used in polar exploration at the turn of the century. Portions of that scarp have an estimated relief of at least 1.24 miles (2 km). Resolution is about 1.2 miles (2 km) in this view. The terrain seen throughout the south polar regions of Mercury resembles heavily cratered regions seen on the incoming hemisphere of the first pass rather than the extensive areas of smooth plains seen on the outgoing hemisphere. Similarly, none of the "peculiar" terrain seen in the first encounter is recognizable in these polar regions. These results confirm the ubiquitous presence of scarps in the older terrains of Mercury and reinforce the conclusion that the scarps reflect a global crustal contraction early in the history of the planet. The absence of any additional occurrences of the peculiar terrain add to its uniqueness and reinforce the preliminary conclusions that associated the formation of that terrain with the Caloris Basin impact, perhaps by focusing of seismic waves at the antipodal point.

127

carrier radio signal from Mariner—we were 'climbing' out of the null! We obtained some 14 [Goldstone's big antenna] coverage and watched for a signal from that spacecraft that would show Canopus. No luck! Time was rapidly running out, so I declared a spacecraft emergency and took 43 [the big antenna in Australia] from Helios. A couple of special analysts were made available to help us figure out the spacecraft's roll rate and its position. They couldn't do it unequivocally—we weren't really sure which direction the spacecraft was rolling in at the time.

"Should I seize 63 [the big antenna in Spain] also? Then what? Another eight hours of staring at the low-gain antenna patterns? If we rolled slowly into a null again there wouldn't be time to reacquire.

"We repositioned the high-gain antenna as a check on our roll direction ambiguity. There was no increase in the received signal, a result consistent with a reverse direction of roll. Finally, having convinced myself by this check and a discussion with the extra analysts that the spacecraft was in a reverse roll, and that the roll rate was low enough to stop safely, I told the operator to send the "stop sequence" command to the spacecraft.

"An agony of suspense followed. If our roll-rate estimate was wrong, the spacecraft would start to oscillate and that would be the end. I'm told that I had been pacing through the mission control facility actually wringing my hands!

"But the spacecraft did stop. Anxiously we looked for the gas pressure telemetry channel data. Just a couple of minutes before the signal was to fade, it produced a barely readable signal! We'd made it!

"We then adopted a suggestion to reacquire Canopus by offsetting the high-gain antenna in such a way that it would lead the spacecraft. We stopped the roll at 40 degrees short of Canopus and then again at 7 degrees short, then reacquired by plotting high-gain-antenna signal strength, and timing the reacquisition command based on the slope of the received automatic gain control curve for which we got a reading from the station on the voice net every 10 seconds. We were using the small antenna at Cebreros, Spain, at this time.

"Three of the team were hunched over a console, plotting these data, with the rest of us hovering over them. The console they were using was disconnected and sitting willy-nilly in an aisle. Actually it was in course of removal to be transferred to the Viking mission. The voice coming over the telephone had a heavy Spanish accent, and we had to ask for repeats every several data points. In retrospect, it was a scene more appropriate to Explorer [the first American Earth satellite] than to Mariner 10.

"The roll error channel climbed toward 128, but the command arrived on time. The spacecraft stopped with Canopus in the acquirable position! Shortly later, we reacquired the star. Only 36 hours before scheduled closest approach of Mariner 10 to Mercury!

"I remember saying to Larry Schumacher, principal architect of the zero gas reacquisition, 'How does it feel after a couple of days of snatching victory out of the jaws of defeat?'"

"I think he was too tired to reply."

March 15, 1975 Only 29 hours before encounter, the TV cameras are ready and all other experiments are "go."

Dr. Norman Ness of Goddard Space Flight Center, principal investigator for the magnetometer experiment, has calculated (from the first Mercury encounter results) what the magnetic field variations to be encountered by Mariner 10 will be like for the new flyby, assuming that Mercury's magnetic field is a scaled-down version of Earth's. As Mariner 10 penetrates the bow shock where the solar wind first meets the planetary effects and is slowed down, the observed time corresponds exactly with the predicted time. Similarly, the penetration of the magnetopause occurs at exactly the right time. The maximum field measured (400 gamma) is slightly less than the predicted maximum because the spacecraft flys by a little higher above Mercury's surface than was expected when Ness made his calculations.

As the spacecraft leaves Mercury, the crossing of the magnetopause and the bow shock again corresponds to predictions. Mercury's magnetic field is unquestionably Earth-like in form (Fig. 6.8). Nature has provided a natural laboratory for study of one of the most intriguing aspects of Earth itself.

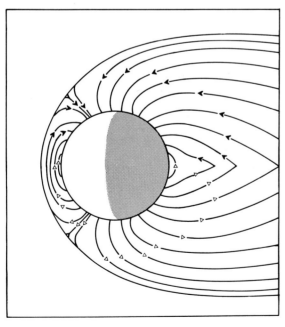

Figure 6.8 The magnetic field of Mercury appears to be a scaled-down version of Earth's magnetic field. This drawing shows a cross-section of the field traced by Mariner 10 as seen from the equatorial plane of Mercury. The arrowheads on the field lines show the direction of the magnetic field. The blunt-nosed boundary to these lines, at left, shows the position of the bow shock where the solar wind is stopped and deflected around the planet. Note how the magnetic field lines are pushed toward the planet at the sunward side (left). North is to the top of the drawing.

129

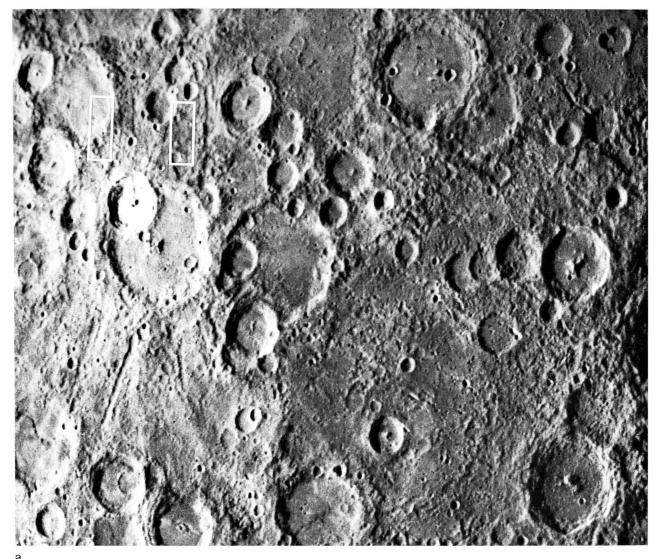

a

Figure 6.9 *THE THIRD TIME AROUND, I.* Figure 6.9a is one of the highest-resolution views of crater Kuiper acquired during the first pass of Mariner 10. The bright rays and ejecta material from the 25-mile (41-km) crater are easily discerned as it nestles on top of an older 50-mile (80-km) crater that also has grooves and other radial ejecta material extending out from it. Indicated by the white rectangles immediately above crater Kuiper are the locations of the views shown in Figures 6.9b and 6.9c, acquired during the third encounter of Mariner 10, on March 16, 1975. The secondary craters formed by projectiles ejected from the primary explosion that produced the crater Kuiper are clearly recognizable by their sharp, fresh character, by their elongation away from crater Kuiper, and by their tendency to occur in chains and multiple groups. The older terrain exhibits a more subdued character, presumably the result of continued impacts and stirring of the surface materials prior to the time of the impact that formed Kuiper. Figures 6.9b and 6.9c were taken from a range of about 6,200 miles (10,000 km) some 18 minutes before Mariner 10 made its third approach. Each frame covers about 44 miles (70 km) in vertical dimension and is displayed at a resolution of about 530 feet (200 m).

130

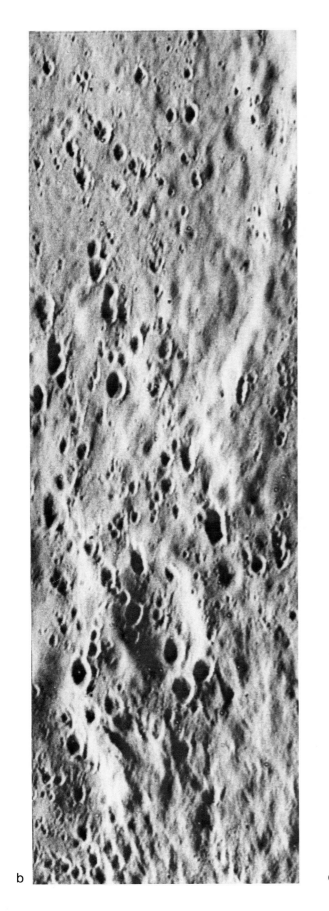

b

c

131

a

Figure 6.10 *THE THIRD TIME AROUND, II.* The third passage of Mariner 10 near Mercury afforded an opportunity to target high-resolution pictures in areas of special interest discovered in the first pass. The peculiar, or hilly and lineated, terrain was given high priority. Figure 6.10a was acquired by Mariner 10 on its first pass, with a vertical dimension of about 310 miles (500 km). White outlines mark the areas of Figures 6.10b and 6.10c, acquired on the third encounter. Figure 6.10b, outlined by the larger box, includes a region of about 93 miles (150 km) in vertical dimension, displayed at a resolution of about 1,300 feet (400 meters). The plains material filling the large crater at the left of the picture is clearly of a younger age than the crater's rim, which has been modified by whatever process caused the formation of hilly and lineated terrain. Figure 6.10c, the right oblong above, includes an area about 43 miles (70 km) long and is displayed at a resolution of about 650 feet (200 m). The area photographed is a very rough terrain with many shadows and slopes facing steeply toward the Sun. In those areas where useful photography can be obtained, small craters are seen down to the limit of resolution, indicating that the surface contains a light crater population similar to the areas of smooth plains on Mercury. At the top of Figure 6.10c is a northwest-trending cliff or slope that has been highly dissected and broken into separate segments. Illumination is from the left in all these pictures.

b

c

133

Figure 6.11 *THE THIRD TIME AROUND, III.* Discovery Scarp was another target
given special priority in the Mercury III passages. Figure 6.11a is a view acquired by
Mariner 10 in the first encounter. Figure 6.11b shows one of the resulting photographs—approximately 106 miles (170 km) in vertical dimension—displayed at a resolution of about 2,300 feet (700 m). The distortion of the craters is especially apparent at
higher resolution, as are their lightly cratered surfaces and the nature of the material filling the craters. In the large crater of the center, a parallel escarpment facing the Discovery is seen, suggesting a small, graben-like structure there. The hexagonal, low-relief
crater at the lower portion of the high-resolution picture on the high side of the Discovery scarp is an anomalous feature, suggestive of volcanic processes although no other
direct evidence in the photograph would require such an interpretation.

a

b

134

a

b

Figure 6.12 *THE THIRD TIME AROUND, IV.* Vostok Scarp, shown in Figure 6.12a as acquired by Mariner 10 in the first encounter, is an especially interesting scarp because it appears to change from a down-to-the-west, sunlit feature in the upper portion to a down-to-the-east, shadowed feature in the lower portion. In the higher-resolution, approximately 0.6-mile (1-km) view in Figure 6.12b, acquired by Mariner 10 on the third encounter, the shortening of the southeast rim of the crater as transected by the scarp is especially apparent. These transection relationships suggest that the scarps correspond to compressional faults, associated with global crustal shortening early in Mercury's history.

a

Figure 6.13 *THE THIRD TIME AROUND, V.* The Caloris Basin was also a high-priority target for photography during the third encounter. Figure 6.13a presents a view acquired during Mercury I. As the view in Figures 6.13b and 6.13c shows, a most interesting pattern of sinuous channels and cracks appears at the maximum resolution acquired. These are possibly similar to collapsed lava tubes and channels postulated to account for some of the sinuous rilles on the Moon, although their general character is distinct from anything seen on the Moon or the Earth. It is possible that a process of flowing of liquid rock on Mercury has operated which has no exact parallel on either Earth or Moon, and is represented by these peculiar channel-like features in the Caloris plains.

b

c

137

The intrinsic Earth-like nature of Mercury's magnetic field is confirmed also by the plasma science experiment, which traces the paths of electrons through the magnetic field. The interaction of these solar wind electrons and the field of Mercury appears as a scaled-down version of what happens in the vicinity of the Earth. The cool and hot regions of the plasma sheet and a polar region of low flux (like similar regions surrounding Earth) reinforce the evidence for an Earth-like magnetosphere of Mercury.

While these non-imaging science experiments are achieving important new results, the imaging experiment, aimed at obtaining high-resolution mosaics of interesting areas, runs into serious difficulties. The encounter had been arranged to take place at a time when data would be returned through Canberra Station of the Deep Space Network, where an experimental ultra-low-noise feed was installed earlier on the antenna. However, as the pass comes up, the low-temperature amplifier of the feed develops a leak in its cooling system. Despite every effort of the engineers, it cannot be repaired in time for the pass. The extraordinary string of luck and maximum success in television imaging from Mariner 10 is ending.

The imaging data have to be returned with the lower bit rate capabilities of the degraded Canberra antenna system. As a result, only a quarter frame of each full picture can be transmitted in the 42 seconds available to send each picture in real time as fast as the cameras aboard Mariner photograph the surface of Mercury. The tape recorder, of course, had failed before the second encounter, so there is no way to store for delayed playback the entire high-resolution frames.

Nevertheless, even strips of high-resolution pictures are important (Fig. 6.9 through 6.13). Disappointed members of the imaging team are reminded that all previous Mariner missions required use of a tape recorder for *any* high-resolution pictures. Had that constraint not been overcome in the development of Mariner 10, there would have been no pictures from either the second or the third encounter and greatly reduced quantities from high-resolution coverage at Venus and Mercury the first time.

March 24, 1975   Mariner 10 uses up the last of its attitude control gas just before noon, barely a week after the third encounter. The spacecraft has traveled a billion miles since leaving Earth 506 days earlier. Its mission is over.

Controllers send the final command to deactivate Mariner. The spacecraft is placed in an all-axis drift mode to see if it can be controlled by solar sailing alone, now that the gas is exhausted. But a short while later Mariner 10 loses both Sun and Canopus references. A command is then sent to turn off the transmitter on the spacecraft. Sadly, its human

138

partners command their robot to sleep. Twenty-one minutes later the Deep Space Network Station receiver loses lock, indicating that the spacecraft's transmitter is dead. Final telemetry signals indicate that the spacecraft has pitched 21 degrees from the Sun. The sunshade no longer shields the vital electronics. Temperatures within the spacecraft are rapidly rising.

Mariner 10, a silent mass of complicated electronics and scientific instruments, pursues an endless journey around the Sun, returning to the vicinity of Mercury every six months. Perhaps future astronauts will retrieve it as a nostalgic act.

Future exploration of the solar system may be by spacecraft taken into space on the space shuttle and carried along by solar sailing.

7

Epilogue

"We work, we think, we explore, we dispute, we take risks and suffer—for there seems no end to the difficult and dangerous adventures individual men and women may attempt; and more and more plain does it become to us that it is not our little selves, but Man the undying who achieves these things through us."
—H.G. Wells, *The Shape of Things to Come* (1929)

What is the scorecard for Mariner 10? How do the actual results compare with the objectives of the Mariner Venus/Mercury project defined in 1969, shortly after the flyby of Mars by Mariners 6 and 7? Mariner 10's primary objective was to fly by planet Mercury making measurements of a complexity comparable to those carried out by Mariners 6 and 7; gravitational assist at the planet Venus was to be utilized in this first attempt at a dual-planet mission. Secondary mission objectives were to acquire useful scientific results at Venus in the course of the flyby there and to gather new measurements of the interplanetary medium within the orbit of Venus, a region of space never before penetrated by a man-made object. This was the first deep-space mission to be attempted at a fixed predetermined total cost. A ceiling was established before submission of cost estimates and competitive bidding by potential industrial contractors. There was no provision in this JPL/NASA fixed-cost agreement for inflation. All of these technical objectives were to be accomplished with but a single spacecraft and at a total cost not to exceed $98 million (exclusive of the launch vehicle).

Mariner 10 flew by Mercury three times and gathered a wealth of scientific information; it returned about 13 times as many useful photographs as did Mariners 6 and 7 together at Mars in 1969. Primary scientific objectives were exceeded. The dual-planet mission was carried out successfully, involving a fourfold increase in navigational accuracy. Eight trajectory correction maneuvers were carried out by Mariner 10 instead of the customary two of previous Mariner missions. In addition, the "solar sailing" technique for conservation of attitude control gas was improvised successfully and thereby qualified as a technique for use in future missions. The use of wide-band radio communications from planetary missions to permit direct transmission of television images without recourse to interim storage in an onboard tape recorder was pioneered successfully on this mission. That technique will be used on subsequent interplanetary missions, such as the Mariner spacecraft to be launched in 1977 for flybys of Jupiter and Saturn.

Two major technological innovations made maximum use of the Venus flyby even though it was of secondary priority. First, the high-gain antenna was programmed for movement about two axes to allow the radio signal to be refracted through the Venusian atmosphere in a teardrop pattern when the spacecraft passed behind Venus. This resulted in much more effective probing of Venus's atmosphere by the radio signals sent from the spacecraft to Earth. Second, the television cameras were reconfigured to permit photography of Venusian clouds in the ultraviolet region of the spectrum, a feature not needed for the observations at Mercury. As a result, otherwise unobservable atmospheric markings on Venus and complex circulation patterns in the upper atmosphere were revealed. The novel wide-band radio communication with Mariner 10 permitted enough "snapshots" of the patterns of these mysterious markings (3,400 individual pictures: 18 times the total pictures returned from Mariner 6 and 7), that global patterns were revealed. This marked a major step in the scientific exploration of Earth's "sister" planet.

The $98 million total cost for Mariner 10 was met—in fact, the project underran by more than $1 million. The extended mission (the second and third encounters) cost an additional $2.75 million, far less than previous experience would have indicated. Even including launch vehicle and indirect costs (which are budgeted separately by NASA), the entire Mariner 10 mission cost about 60 cents total per American during that five-year period—barely the price of a single gallon of gas! And, of course, those costs represented mostly the salaries of individual Americans carrying out challenging tasks with dedication, increasing their capability for future tasks as well.

The Mariner Venus/Mercury mission not only proved the feasibility of the fixed-price approach but also successfully employed a "cost incentive" contract mechanism with the spacecraft system contractor, the Boeing Company. Under this arrangement, a significant award fee was an incentive for satisfactory technical compliance within cost limits. Boeing was, indeed, successful in this regard, and ultimately was awarded a fee of $5 million on a total effort of $43 million. This handsome profit was a well-deserved reward for success in implementing a challenging technical task at less than the negotiated target cost. This managerial achievement is even more significant in view of the fact that during the period of the Mariner 10 mission, inflation was eroding the dollars available to accomplish the work—and no provision was made in the project's fixed-price agreement to accommodate that inflation. Whereas in earlier days aerospace contractors sometimes seemed to profit from cost overruns on ambitious projects, Mariner Venus/Mercury proved that profit for the aerospace contractor could be tied to economical and efficient implementation of a difficult technological endeavor for the federal government.

Mariner 10 was thus an unqualified scientific, engineering, and managerial [1] success, and its accomplishments have been recognized formally with many awards and commendations (see Appendix B). Another highly valuable and enduring reward is the individual feeling of personal accomplishment by those who made it possible. Rare indeed is the opportunity to be challenged to the limits of one's capabilities by a positive, creative task and to discover unsuspected inner resources while meeting that challenge.

What has happened to the participants in the Mariner Venus/Mercury mission, those modern revolutionaries whose historic role is belied by their conservative appearance? Some are incurable in their quest for new challenges. Without even time to catch his breath, Gene Giberson left behind the hard-won success of Mariner 10 to take over leadership of a new flight project intended to demonstrate the practical value of novel techniques for surveying the oceans from Earth-orbit. The SEASAT project, like Mariner Venus/Mercury, is a difficult technical endeavor caught in a cost squeeze from the beginning. The problem should seem familiar—if formidable—to Gene Giberson and to Jim Dunne, and to half a dozen others who, in time, followed Giberson's lead to SEASAT.

Indeed, well over half of the hundred or so engineers and scientists who worked on Mariner Venus/Mercury at the Jet Propulsion Laboratory have gone on to other flight projects, including the Viking mission to orbit and land on Mars in 1976 and, especially, the Mariner Jupiter/Saturn mission scheduled for launch in 1977 and arrival at Jupiter in 1979 and at Saturn in 1981. A few, like Ed Danielson, even managed to become involved in the Mariner Jupiter/Saturn project right from the

---

[1] The Mariner Venus/Mercury management team was recognized by the Presidential Management Improvement Award in 1974, one of only 12 endeavors in the entire country to be so designated.

beginning despite their heavy responsibilities for Mariner 10, and thus shaped it a little according to their Mariner 10 experience.

John Casani's success in managing the Mariner 10 spacecraft development (on top of previous successes with Mariners 4, 6, and 7) resulted in his selection for a major technical management position at JPL, manager of the Guidance and Control Division. Similarly, Daryal Gant, a key man in the complex and successful contractual relations and procurements associated with Mariner 10, was selected to head the Procurement Division, a major administrative unit at JPL. Bill Cunningham, the key official in NASA's Washington, D.C., headquarters for Mariner 10 (and previously in the same capacity for Mariners 6 and 7 and the early Ranger series of TV Moon probes), now is putting his hard-won experience to good use in NASA's new Low Cost Systems Office.

Bruce Murray had planned for Mariner Venus/Mercury to be his last flight project after intensive involvement encompassing the Mariners 4, 6 and 7 Mars flybys as well as the Mariner 9 Mars orbiter. "Four times around that obstacle course is enough for anyone!" A Guggenheim Fellowship in the pleasing environment at La Jolla provided a perfect sequel to Mariner 10, with an opportunity to contemplate the basic significance of the enormous recent discoveries concerning the Earth's planetary neighbors. Afterward, however, instead of returning to a moderated pace of teaching and quiet research, he was selected to succeed Dr. Pickering, who was retiring as Director of JPL.

Ed Czarnecki, Boeing's top man on Mariner 10, was, like Giberson, off and running on a new project (hydrofoil development) before Mariner 10 had even completed its mission. The aggressive efficiency that characterized Boeing's building of the Mariner 10 spacecraft also led to a rapid reassignment of some of the engineering force to other jobs; for others, there were layoffs. The aerospace industry is notoriously unreliable as an employer; employment depends on new business, not yesterday's fantastic success.

Not all the Mariner 10 veterans at JPL slipped easily into challenging new assignments. Vic Clarke, project gadfly *par excellence*, found that few really challenging new endeavors were being sponsored by the federal government in the psychologically and economically depressed times following Mariner 10's odyssey. He and some other exceptionally creative people had to struggle with a succession of diverse studies and analyses concerned with future possibilities in space, as well as in important domestic technological areas, before the right combination of imaginative federal technological initiatives and personal good fortune once again provided the opportunity for special individual achievement like that represented by the mission to Mercury.

And what about Giuseppe Colombo, that latter-day Columbus who helped navigate to a twentieth-century new world? He continues to try to substitute brain-power for rocket power to return another spacecraft to Mercury. With like-minded collaborators at JPL and at the Smithsonian Institution in Cambridge, Massachusetts, he is investigating even more exquisite celestial billiard shots to Mercury by way of Venus. ("And what if the rocket motor is fired *just at the time of closest passage by Venus on the way to Mercury?*") If there is to be a new spacecraft returning to Mercury in the near future, if there is to be any continuity in the exploration of Mercury initiated by Mariner 10, it will require some real inventiveness to get there cheaply.

What was learned in this revolutionary campaign? What is the meaning of the thousands of pictures [2] and reels of magnetic tape full of other kinds of scientific data? The scientists directly involved with Mariner 10 have published their findings in widely read scientific journals. More popular accounts have appeared in *Scientific American* and *National Geographic.* [3] Finely detailed airbrushed maps showing the topographic features of Mercury are being produced by the United States Geological Survey, and a superbly illustrated and reproduced atlas of Mercury pictures will be available soon.

Science is a continuing human process; perceptions change as new data are acquired and as new ideas suggest new meaning for old data. The enormous intellectual impulse of the Mariner 10 observations has not only revolutionized attitudes about the planet Mercury but also raised new questions about the other terrestrial planets, including Earth itself. Two hundred and twenty-five scientists from many nations met in Pasadena in June 1975 at the First International Colloquium on Mercury to discuss and debate the findings of Mariner 10. Mercury is now viewed as being rather Earth-like inside, yet seemingly it records a Moon-like surface history. But why should Mercury's surface history, the aggregate scars of external bombardments and blemishes of volcanic and tectonic activity, resemble that of the Moon? Did both planetary bodies experience similar early histories of impact by asteroid-sized bodies despite their very different solar-system habitats? Did both bodies experience similar early episodes of widespread vulcanism, followed by inactivity for billions of years, despite their very different internal constitutions? (See Fig. 7.1 through 7.3.) None of these questions can yet be answered to every specialist's satisfaction, as was apparent from heated debate at the Colloquium.

What has become quite probable is that heavy bombardment took place

[2] The original picture data are publicly available through the National Space Science Data Center, Goddard Space Flight Center, Greenbelt, Maryland.

[3] See July 12, 1974, issue of *Science;* June 10, 1975, issue of *Journal of Geophysical Research.* More popular presentations can be found in the September, 1975, issue of *Scientific American,* the May, 1975, issue of *National Geographic,* and the December, 1975, issue of *Astronomy.*

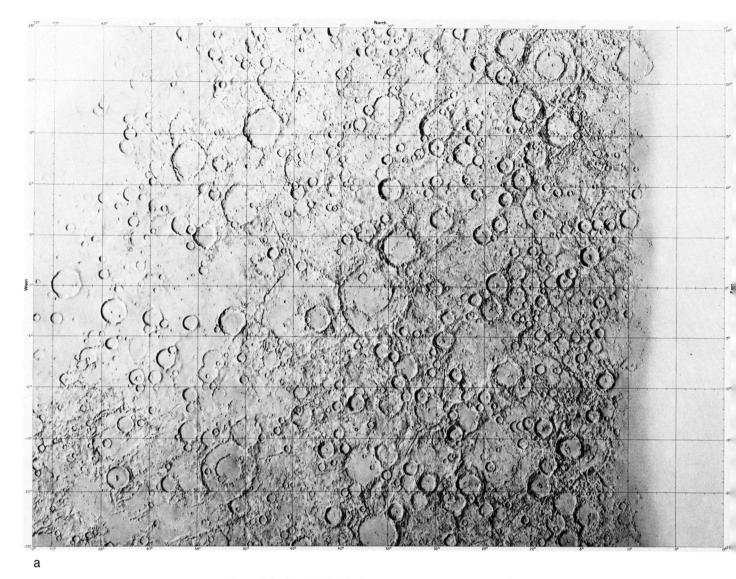

a

Figure 7.1 *THE REFERENCE MERIDIAN OF MERCURY.*

Figure 7.1a.   Airbrushed interpretation of the Kuiper
Quadrangle of Mercury, produced by the United States
Geological Survey. This version shows topographic
features but suppresses albedo markings,
i.e., differences in light and dark colorations.

Figure 7.1b.   Enlarged section of Kuiper Quadrangle
around the equator and twentieth (reference) meridian.
The black box outlines the area covered by the single
Mariner television image of Figure 7.1c.

Figure 7.1c.   Mariner image showing crater Hun Kal,
which defines meridian 20°W. The parallels are defined
by the rotational axis, currently assumed to parallel the
orbital axis. The choice of the name Hun Kal seems especially apt, as it signifies the number 20 in the language of
the Mayans, whose excellent astronomical observations
used a number system based on 20 rather than 10 as
does the Arabic system we use.

146

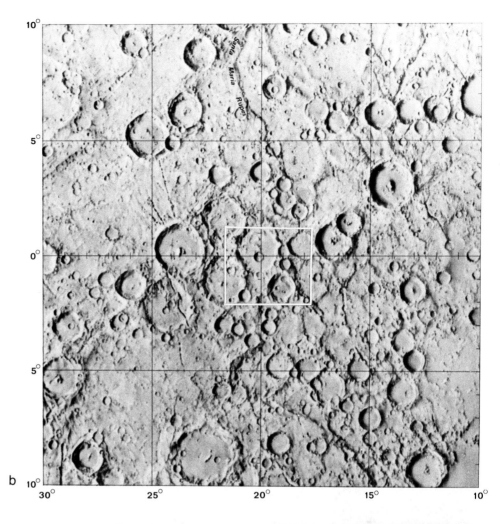

b

c

HUN KAL

Figure 7.2 *MERCURY'S CORE EXPOSED.* The core is drawn under a photomosaic of Mercury from the approach side. The size of the core, much larger proportionately than that of Earth, is determined from knowledge of the average density of the planet and the assumption that it is principally composed of the same elements as Earth.

148

on all the terrestrial planets as late as half a billion years after they first began to accrete into planets. This late bombardment seems to have been a period of degradation and erosion of the planetary surfaces quite distinct in nature from the initial accumulation of the planets, which also involved collision with small objects but under conditions conducive to accumulation rather than erosion. We know from laboratory dating of lunar samples brought back by the Apollo astronauts that the lunar surface experienced heavy bombardment up to about 4 billion years ago even though original accumulation probably took place in a brief interval about 4.5 billion years ago. Four to 4.5 billion years ago is a very remote time period—even from a geological point of view—but highly significant because the oldest rocks still preserved in isolated localities on Earth are just a little younger, 3.7 to 3.9 billion years old. Thus, this previously unrecognized terminal bombardment episode of the inner solar system, which surely involved Earth as well, immediately predates the surviving geological record here. Perhaps that bombardment destroyed earlier records on Earth, as it has on the Moon. Furthermore, even the oldest surviving geological records on Earth seem to record surface conditions much more like present ones than like those that must have accompanied initial accumulation. The ocean had already come into being and the atmosphere even then probably resembled the present one in most respects. Life had begun, and blue-green algae (at least) had evolved. Evidently, much of Earth's planetary evolution took place earlier than represented by surviving records.

In reaching out to explore his planetary neighbors and his Moon, man has found otherwise unobtainable records of bizarre early planetary conditions that must have profoundly affected Earth itself (and its biological evolution down to man) in its formative stages. Comparison of the lunar and Mercurian records (and the rich record from Mars as well) greatly increases the independent significance of each and dramatically illustrates the multiplicative, rather than merely additive effect of new scientific knowledge.

In the late nineteenth century, geologists first began to compare the geological records from different continents; from such endeavors the global history of the Earth first began to be distinguished from local events and processes. In the latter part of the twentieth century, the new technology of spaceflight—and the optimism and enthusiasm of the American people—have made it possible to begin to correlate the records from different planets, thereby helping to sort out common elements of planetary history from conditions and events unique to each body.

The multiplicative value of planetary exploration was illustrated in another way by the Mercury findings of Mariner 10. The origin of the Earth's magnetic field has challenged theorists since Sir William Gilbert proposed, in 1600, that the Earth acts as a permanent magnet. In modern times, scientists have generally believed that the core of the Earth is electrically quite conductive because it is composed mainly of very hot iron and nickel that flows slowly in a manner crudely analogous to that of a highly viscous heated liquid (i.e., in convection).

A differentially moving electrical conductor that is also rotating as a whole, due to the Earth's spin, can interact with itself to produce a magnetic field, at least hypothetically. This dynamo concept can explain much of the observed behavior of the Earth's magnetic field, such as its surface intensity and orientation and temporal changes, *if* plausible guesses are made concerning a host of specific properties of and conditions within the core. Obviously, there is no way to measure directly the properties of the materials 2,000 miles and more inside the Earth. Analysis of earthquake seismic waves and other geophysical measurements, provides limits on some combinations of the physical properties. But the theoretical model is sufficiently unconstrained by facts that it could just as easily explain a somewhat different magnetic field at Earth's surface by using some equally plausible guesses at the properties of the core.

The situation seemed destined to remain ambiguous indefinitely—until Mariner 10's surprising discovery of an Earth-like magnetic field at Mercury. Nature, it appears, has been generous enough to provide a second, probably simpler, example of an Earth-like field. Dynamo models of the Earth's field, when extrapolated to a planet of Mercury's slow rotation, small size, and likely interior conditions, must now predict a magnetic field like that actually observed there. Furthermore, Venus, despite its Earth-like size, mass, and probable internal conditions, does not exhibit a magnetic field. Thus, a dynamo theory adjusted to explain Earth and Mercury must predict as well the absence of a reasonable field on Venus, perhaps because Venus's rotation rate is even slower than Mercury's (once in 243 days versus once in 58.5 days). Measurements at Venus and especially Mercury are therefore directly pertinent to an otherwise dead-ended terrestrial scientific problem.

Systematic mapping of the Mercurian magnetic field from a future orbiting spacecraft would be of significance to pinning down the origin of Earth's field. The same orbiter could, of course, map photographically the as-yet-unviewed hemisphere of Mercury, which might contain new (or at least clearer) geological records, as well as make other measurements to determine a good deal more about the internal constitution and surface composition. Going back to Mercury would help to elucidate aspects of our own planet that are buried too deep in space or time for direct investigation. And who knows what unanticipated discoveries await man's next, more detailed, scrutiny of the strange little planet nearest the Sun?

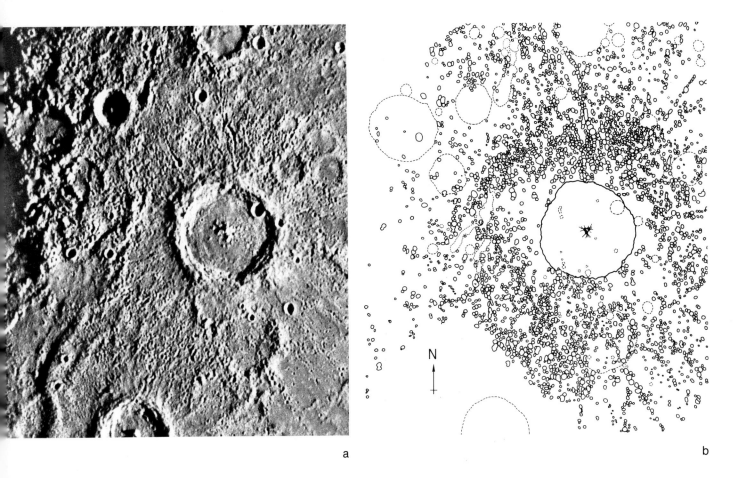

a   b

Figure 7.3 *WHAT GOES UP MUST COME DOWN!* Secondary crater field associated
with large (140-km diameter) Mercurian crater. The relatively tight pattern of ejecta from
the original impact is characteristic of such craters on Mercury. To aid in interpretation,
the original image has been geometrically transformed to an orthogonal projection by
computer.

Suppose, however, that refined observations of the Mercurian magnetic
field, and perhaps better upper limits for Venus, prove to be inconsis-
tent with *any* reasonable extrapolation of the dynamo theory of the
Earth. Another, perhaps unimagined, explanation for the Earth's field
might be indicated. Indeed, it is probably only through careful comparison of Mercury's and Earth's magnetic fields that the dynamo theory can
really be tested. Otherwise, it is difficult to see how scientists could ever
proceed beyond the domain of a "plausible explanation," sufficient to
dull the quest for alternatives but never really provable unambiguously.

The study of the Earth is becoming a part of a broader intellectual
framework: comparative study of the terrestrial planets. Mariner 10, be-

151

sides providing a twentieth-century outlet for *Homo sapiens'* fundamental curiosity about his environment, has proved to be a powerful instrument for major scientific discovery pertinent to the nature and history of our own planet.

So much for Mariner 10, its results, and the people who made it happen. The violence, upheaval, and change on Earth accelerate as humanity rushes toward the challenge of the future. The fall of Saigon in early 1975 culminated the crescendo of domestic and international change that counterpointed Mariner 10's every stage. That final failure may in some ways have denoted the end of an extraordinary era for the United States in which technology was so successfully used for high achievement with Apollo and so unsuccessfully used as a bloody substitute for political leadership and international wisdom in Southeast Asia. Mariner 10 was conceived near the end of that era and in many ways was a late spin-off of the Apollo enthusiasm and drive.

Have we entered a new and qualitatively different era? Few new, exciting endeavors of any kind are being started by an inward-looking federal government. No new United States projects have been authorized to explore the planets since 1973. Apollo and Vietnam alike arose from the great passions born of Soviet and American rivalry in the early 1960s. Whence will come new impulses to unite America in search of the great accomplishments to draw from us unique national contributions to the world's history? Americans have not easily accepted a dominant rather than predominant role in world relations. They remain frustrated by the continuing disparity between ideals and reality within their own country; the lofty vision of the former seemingly demeans the historic achievements represented by the latter. But America has such diversified talent and resources that it can, if it so wills, achieve high goals even as it struggles to find new and more effective means to deal with both traditional and newly created forms of social disharmony and dissatisfaction.

The story of Mariner 10 is an important reminder of our opportunities and values. Mercury has been made a real part of the harassed and over-stimulated twentieth-century consciousness. The revolution of ideas continues. The mind of man has been permanently altered by the pictures and other data returned by Mariner 10 to constitute an enduring and culturally valuable legacy.

Appendix A

MARINER VENUS/MERCURY 1973 SCIENCE INVESTIGATORS

Television Experiment

TEAM LEADER   Bruce C. Murray
California Institute of Technology

TEAM MEMBERS   Michael J. S. Belton
Kitt Peak National Observatory

G. Edward Danielson, Jr.
Jet Propulsion Laboratory

Merton E. Davies
Rand Corporation

Donald E. Gault
NASA Ames Research Center

Bruce Hapke
University of Pittsburgh

Brian T. O'Leary
Hampshire College

Robert Strom
University of Arizona

Vernon E. Suomi
University of Wisconsin

153

Newell J. Trask
U.S. Geological Survey

ASSOCIATE TEAM MEMBERS James L. Anderson
California Institute of Technology

Audoin Dollfus
Observatoire de Paris

John Guest
University of London Observatory

Robert Krauss
University of Wisconsin

Plasma Science Experiment

PRINCIPAL INVESTIGATOR Herbert S. Bridge
Massachusetts Institute of Technology

CO-INVESTIGATORS J. Asbridge
Samuel J. Bame
Los Alamos Scientific Laboratory

W. C. Feldman
Los Alamos Scientific Laboratory

A. Hundhausen
University of Colorado

Leonard Burlaga
R. E. Hartle
Keith W. Ogilvie
J. D. Scudder
NASA Goddard Space Flight Center

J. H. Binsack
A. J. Lazarus
S. Olbert
Massachusetts Institute of Technology

George L. Siscoe
University of California at Los Angeles

154

Clayne M. Yeates
Jet Propulsion Laboratory

Ultraviolet Spectroscopy

PRINCIPAL INVESTIGATOR   A. Lyle Broadfoot
Kitt Peak National Observatory

CO-INVESTIGATORS   Michael J. S. Belton
Kitt Peak National Observatory

S. Kumar
Jet Propulsion Laboratory

M. B. McElroy
Harvard University

Infrared Radiometry

PRINCIPAL INVESTIGATOR   Stillman C. Chase, Jr.
Santa Barbara Research Center

CO-INVESTIGATORS   Ellis D. Miner
Jet Propulsion Laboratory

David Morrison
University of Hawaii

Gerry Neugebauer
California Institute of Technology

Charged Particle Experiment

PRINCIPAL INVESTIGATOR   John A. Simpson
University of Chicago

CO-INVESTIGATOR   J. E. Lamport
University of Chicago

155

Radio Science Experiment

TEAM LEADER   H. T. Howard
Stanford University

TEAM MEMBERS   John D. Anderson
Gunnar Fjeldbo
Arvydas J. Kliore
Gerald S. Levy
Jet Propulsion Laboratory

Irwin I. Shapiro
Massachusetts Institute of Technology

ASSOCIATE TEAM MEMBERS   D. Lee Brunn
Richard Dickinson
Pasquale B. Esposito
Warren L. Martin
Charles T. Stelzried
Jet Propulsion Laboratory

R. D. Reasenberg
Massachusetts Institute of Technology

G. Tyler
Stanford University

Magnetic Fields

PRINCIPAL INVESTIGATOR   Norman F. Ness
NASA Goddard Space Flight Center

CO-INVESTIGATORS   Kenneth W. Behannon
Ronald P. Lepping
Kenneth H. Schatten
NASA Goddard Space Flight Center

Y. C. Whang
Catholic University

Appendix B

THE 1974 MARINER 10 NASA AWARD RECIPIENTS

Public Service Group Achievement Awards

Boeing Aerospace Management Team
Data Records Group
Science Instrument Development Team
Television Science Team
Spacecraft System Design Team
Temperature Control Design Team
Boeing Cognizant Work Unit Engineers

Group Achievement Awards

Flight Project Representative Team
Flight Data Subsystem Development Team
Ground Data System Integration Team
Mission Control and Computing Center
Mission Sequence Working Group
Navigation Development and Operations Team
Roll Axis Anomaly/Solar Sailing Team
Television Subsystem Development Team
Work Unit Management Team
System Contract Procurement
Spacecraft Flight Operations and Mission Control Teams
Mariner 10 Headquarters Staff Support Group
Jet Propulsion Laboratory

Public Service Awards

Richard A. Axell
William E. Bramel
Kunihei Kawasaki

Haim Kennet
Bernard M. Lehv
George B. Rickey

Exceptional Service Medals

Lida M. Bates
Lyle V. Burden
Elliott Cutting
G. Edward Danielson, Jr.
Esker K. Davis
Richard L. Foster
Daryal T. Gant
Harold J. Gordon
Adrian J. Hooke
William R. Howard (deceased)

Edward H. Kopf, Jr.
William I. Purdy, Jr.
Norri Sirri
F. Louis Sola
Anthony J. Spear
Gael F. Squibb
Francis M. Sturms, Jr.
Fred Vescelus
Peter B. Whitehead
James N. Wilson

Exceptional Scientific Achievement Medals

Herbert S. Bridge
Victor C. Clarke, Jr.

James A. Dunne
John A. Simpson

Distinguished Public Service Medals

Bruce C. Murray
Edwin G. Czarnecki

Outstanding Leadership Medal

John R. Casani

Distinguished Service Medal

Walker E. Giberson

158

Index

161